SCIENTIFIC INSTRUMENTS
1500-1900

SCIENTIFIC INSTRUMENTS
1500-1900
An Introduction

Gerard L'E Turner

PHILIP WILSON

———

University of California Press

Berkeley   Los Angeles   London

BY THE SAME AUTHOR:

Descriptive Catalogue of Van Marum's Scientific Instruments in Teyler's Museum (1973); *Essays on the History of the Microscope* (1980); *Antieke wetenschappelijke instrumenten* (1981); *Historische microscopen* (1981); *Nineteenth-Century Scientific Instruments* (1983); *The Great Age of the Microscope: The Collection of the Royal Microscopical Society through 150 Years* (1989); *God Bless the Microscope! A History of the Royal Microscopical Society over 150 Years* (1989); *Scientific Instruments and Experimental Philosophy 1550–1850* (1991); *Museo di Storia della Scienza: Catalogue of Microscopes* (1991); *Storia delle scienze I: Gli strumenti* (1991); *The Practice of Science in the Nineteenth Century: Teaching and Research Apparatus in the Teyler Museum* (1996)

ACKNOWLEDGEMENTS AND PICTURE CREDITS

I thank my colleagues in the University of London and the University of Oxford, and my friends in many museums all over the world for their co-operation in my study of the history of scientific instruments over a period of 30 years. I gratefully acknowledge financial support over the years for travel from the Wellcome Trust, the Nuffield Foundation, the Renaissance Trust, the Leverhulme Trust, the Royal Society, the University of Oxford, and for a visit to Italy a NATO Collaborative Research Grant.

I acknowledge with gratitude the provision of photographs by the following (key words are in the captions).

Christie's Christie's South Kensington, London

GLET   Author's photograph

MHSG   Musée d'Histoire des Sciences, Geneva

MHSO   Museum of the History of Science, Oxford
NMM   National Maritime Museum, Greenwich

PC   Private Collections

PU   Department of Physics, University of Pavia, Italy

RMS   Royal Microscopical Society at the Museum of the History of Science, Oxford

SA   Soroe Academy, Soroe, Denmark

SM   Science Museum, London

Sotheby's Sotheby's, New Bond Street, London

TM   Teyler Museum, Haarlem, The Netherlands

WMHS   Whipple Museum of the History of Science, Cambridge

This edition published 1998 by
Philip Wilson Publishers,
143-149 Great Portland Street,
LONDON W1N 5FB

ISBN 0 85667 491 5

University of California Press
Berkeley and Los Angeles, California

University of California Press Ltd.
London, England

US ISBN 0-520-21728-4

First published 1980 as *Antique Scientific Instruments*

Editors: John Gilbert, Tim Ayers
Design: Peter Ling

Colour origination by Saxon Photolitho, Norwich
Printed and bound in Italy by Artegrafica S.P.A.-Verona

9   8   7   6   5   4   3   2   1

Front cover illustration: A collection of scientific instruments, dating from *c.* 1740 to *c.* 1870. (Photo courtesy Peter Delehar)

Colour illustration on p. 2: Wimshurst pattern electrostatic induction generator, with four Leyden jars and adjustable spark gap at the top; signed: *Harvey & Peek, 56, Charing Cross Road, London (Late W. Ladd & Co.)*; *c.* 1890. (Christie's)

Colour illustration on p. 3: English gilt brass pocket equinoctial dial with magnetic compass and nocturnal, signed: *Elias Allen fecit*; *c.* 1630. (Christie's)

CONTENTS

A typical pair of English 12-inch (30.5 cm) diameter globes made in the first half of the 19th century. The celestial on the left is dated 1816, while the terrestrial is dated 1844. Both were made by J. & W. Cary, London. (Christie's).

INTRODUCTION

AFTER the classical period, astronomy and horology were studied in the Islamic communities that occupied the east, south and west of the Mediterranean. Some scientific knowledge infiltrated into Western Europe via Spain, chiefly in the fields of alchemy and astrology. Although such studies find no place in modern science, it is the theorising that is unacceptable; the instruments and manipulative skills necessary for measuring and provoking phenomena are neutral. An astrolabe which incorporates a means for measuring the elevation angles of stars or the Sun, as well as a stereographic representation of the Earth, can be used equally well for telling the time, finding the latitude, and for collecting data to cast a horoscope.

The Italian Renaissance laid the foundations of modern science that eventually freed itself from the superstition inherent in alchemy and astrology. The discovery of the American continent by Columbus in 1492 began an economic rivalry among European nations that required for the first time astronomical navigation with the aid of instruments. The publication, in 1543, of *De revolutionibus orbium coelestium* by Nicolas

Copernicus changed thinking about the universe, with the realisation that the planetary system was simpler to describe if the Sun and not the Earth were regarded as central. These revolutionary new ideas resulted in a demand for increasing accuracy in observational astronomy, pioneered by the Dane, Tycho Brahe. As trade with the New World and between Europe and the East developed in the 16th and 17th centuries, so did the need for a reliable means to establish a ship's position when crossing the oceans. This led to the publication of manuals on the mathematics of navigation, and to the improvement of such instruments as the mariner's astrolabe, the quadrant and the compass. The problem of finding the longitude out of sight of land was understood in principle, but could not be solved because there was no mechanism capable of being carried on board ship that would keep the time at the port of departure. Christiaan Huygens, in the 1660s, attempted to make a sea-going pendulum clock, but the solution had to wait for the invention by John Harrison of the marine chronometer in the mid-18th century.

Exploration, and the need to define the boundaries of property, required the development of instruments for surveying, beginning in the mid-16th century with the invention of the forerunner of the theodolite by Leonard Digges. The instruments used for astronomy, time-telling, navigation and surveying are known collectively as mathematical instruments. They were so called because their purpose was to make measurements, and because a mathematical principle was embodied in their construction.

Making a sundial, for example, presents a trigonometrical problem, and its purpose is to measure and record the passing of time, one of humanity's most basic needs. The astrolabe was an omnibus instrument, which told the time, and also made possible angular measurement for both astronomical and surveying purposes. All physical measurement basically reduces itself to measuring time, length and mass. Other mathematical instruments include the quadrant, for measuring elevation, the theodolite for surveying, and a wide variety of rules, graduated scales and devices for recording weight and volume. The makers of these instruments were men who not only had skill in metalwork and engraving, but also mathematical knowledge. The invention of the printing press in the mid-15th century, as well as the economic impetus of the voyages of discovery, created the map and chart trade in Flanders, where many of the instrument makers learned their skill.

Given that the properties of mirrors were known from classical times, and spectacles were in use in the Middle Ages, it is strange that optical instruments were comparatively late to develop. The combining of two lenses together was not seriously attempted until the beginning of the 17th century. The invention of the telescope must be credited to Dutch spectacle-makers in 1608; among the earliest users were Galileo, for his famous observations of the moons of Jupiter in 1610, and Thomas Harriot in London. One reason for the late development of optical instruments was undoubtedly the poor quality of glass available. This inhibited the performance of both telescopes and microscopes until the 19th century, and led to the invention of the reflecting telescope, in which the optics are made not of glass but of polished metal. Yet even allowing for these problems, optical instruments added to human experience the new dimensions of the very distant and the very small, producing an enormous popular impact. Among the most striking of the images made available by the microscope was the circulation of the blood, the theory of which was first propounded by William Harvey in 1628. Thereafter, for more than two centuries, microscopes were supplied with a fish-plate to hold a specimen so that the blood-flow in the tail fin capillaries could be observed.

The 16th-century scholar, Francis Bacon, contributed so much to the change in scientific method that he may be hailed as the herald of modern experimental science. He taught that an explanation or theory that has not been tested is of no value, because it lacks a basis in the real world, the yardstick by which scientific truth is recognised. The Royal Society of London originated as a consequence of a meeting at Gresham College in 1660 when it was decided to found 'a Colledge for the promoting of Physico-Mathematicall Experimentall Learning'. The Royal Society early appointed a Curator of experiments and adopted as its motto *Nullius in verba*. The teaching of science by experiment and demonstration spread across Europe, from the universities to learned societies and clubs, social gatherings, coffee-houses and rooms hired by lecturers who travelled the country with their apparatus.

Scientific instruments were divided by the makers themselves into three broad categories. The man who sold the full range of instruments would describe himself as 'A Mathematical, Optical and Philosophical Instrument Maker', though there were some specialists, such as James Short, who made 'solely reflecting telescopes'. The philosophical instruments were those pieces of apparatus designed to produce and demonstrate various physical effects. Both the air-pump and the frictional electrical machine were popular.

The air-pump of von Guericke, designed to create a vacuum, was redesigned in 1658 by Robert Hooke, later Curator of experiments to the Royal Society of London, while he was assistant to Robert Boyle in his research on gases at Oxford. There was much controversy throughout the 18th century on the nature of electricity, and electrical machines were used both by serious scientists and for spectacular popular demonstrations. Electricity was claimed to have a therapeutic effect, and the Reverend John Wesley believed it to be a heaven-sent source of cheap medical treatment for the poor.

There were many other demonstration pieces of apparatus, designed to show the effects of magnetism, heat, gravity, friction, and so on. These instruments continued to be produced for teaching purposes throughout the 18th and 19th centuries, and some have found a new lease of life today as playthings.

The impulse to collect is almost universal, satisfying the hunting and acquisitive instincts, the love of beauty and intellectual curiosity. The wealthy have collected rare and beautiful things from the earliest days of civilisation, but the collection, or 'cabinet', containing natural curiosities, dates from the 16th century. This type of collection also accommodated scientific instruments. Superbly crafted and embellished instruments were made for princely collectors by such makers as Erasmus Habermel, Gerard Mercator and Walter Arsenius. Later, as the interest in science spread through society, smaller collections were made for use by societies, and in the home, where science could provide popular recreation.

In the 20th century, we have come to accept a vast range of technical, often very complex, equipment for everyday use. Science has become the very substance of our lifestyle. Nevertheless, the appeal of historic scientific instruments remains, and from them much can be learned of the practice and development of science over four centuries.

An English horizontal dial signed and dated: H. Cole 1579; made for Sir Henry Darcy. Humfrey Cole was a fine Elizabethan instrument maker working in the period 1568 to 1590. (MHSO)

ASTRONOMY AND TIME-TELLING

P EOPLE have marvelled since the beginning of recorded time at the starry sky; watching and interpreting it has always been a fundamental instinct. To impose order on what was seen, star groups in the form of animals, gods and heroes were identified as reference points. The interpretation of the heavens duly led to two separate strands of enquiry: astrology and astronomy. Most cultures have sought to discover in the stars laws that govern the human mind and body, explaining the origin and foretelling the future of mankind. This is the world of the astrologer, whose task is divination, attempting to satisfy man's perennial longing to look into the future.

The astronomer, however, engages in the systematic study of the heavens for a variety of practical reasons: for telling the time by day and night, for guidance in travel by land and sea, and for recording the progress of the seasons. No civilisation could function without a calendar. Astronomy is thus the most ancient of the sciences, and from earliest times has required the use of instruments for measurement and observation.

The first observatories were monumental in scale, using natural features allied to stone pillars, such as are to be found in Brittany. It has been proposed that Stonehenge was a solar and lunar observatory. There was a strong Islamic tradition of astronomy that linked with that of the Greeks after Ptolemy's *Almagest* was translated into Arabic. In the early 15th century, Ulugh Beg, prince and astronomer, built a monumental observatory in his city of Samarkand, and also produced the first star catalogue since Ptolemy. The Indian ruler Sawai Jai Singh II, much influenced by Ulugh Beg, built a number of observatories in the 18th century, the largest at Delhi and at Jaipur, where they still survive (**13**).

In Europe, the first observatory instruments to achieve serious accuracy were those devised by Tycho Brahe, the Danish astronomer, at the end of the 16th century. Johannes Kepler used these for the calculations that led him to discover the laws of planetary motion. From the mid-17th century, observatories were founded in most European countries, and ever-increasing accuracy was achieved by instrument makers in providing the large, often wall-mounted apparatus used by astronomers.

The time-telling instruments described in this book, however, are those that were portable. These increased in number and variety with the need to tell the time ever more accurately, and served a wide range of personal and professional requirements. In the 15th and 16th centuries, most learned men would have owned an astrolabe, while the sundial in one of its many forms was as ubiquitous as the watch is today.

Note on Hours

The development of modern communications has made the whole world accept the 24-hour clock, although many clocks and watches show a period of 12 hours only. The division of the day into double 12 or 24 parts is exceedingly old, although the day and night were not always divided as they are now. It is necessary to understand which system was used in order to interpret the lines and marks on the astrolabe, quadrant and sundial.

Planetary Hours, so called because astrologers supposed each hour to be ruled by a planet, were also named 'unequal hours', 'temporal hours' or 'Jewish hours'. Here the periods of daylight and darkness are each divided into 12 hours, the key times being sunrise and sunset. This means that midday and midnight are the sixth hour, and that the day and night hours are not of the same length. They can be equal only at the Vernal and Autumnal Equinoxes; at midsummer in northern Europe, the night-hour could equal 40 minutes of our time, and the day-hour 80 minutes of our time. At midwinter, of course, these times would be reversed, with 80 for night and 40 for day.

Babylonian Hours, also 'Greek hours', belong to a system where the day plus night is

divided into 24 equal hours, but they are numbered from sunrise to the next sunrise. Thus, the middle of the day would be at a different numbered hour as the year progressed. Italian Hours are similar to Babylonian hours, but numbered from sunset to the next sunset.

Uniform Hours take their name from their standard length. By the 14th century clocks with bells, and then with faces, began to be placed in churches and other important buildings; and since a mechanical clock must be regular in its movement, a system of hours uniform in length came into general use, numbered from 1 to 12 starting from midday or midnight. The rate of the clock, its accuracy, had to be checked by a sundial, and this necessitated new designs that would supersede the scratch dials incised on church walls by the Saxons and their successors.

Astrolabe

In many ways this is the archetypal scientific instrument, because of its antiquity and remarkable sophistication (1–4). The astrolabe (the name means 'star-taking') is a flat, circular instrument, usually of brass, occasionally of silver, which embodies a stereographic projection of the globe, and of the hemisphere of the heavens; a sort of flattened armillary sphere. The point of projection is nearly always the South Pole, and the plane of the projection is the Equator. This is why scholars refer to the instrument as a planispheric astrolabe, in distinction to the spherical astrolabe, which is a model of the globe and the heavens, needing no lines of projection. Only one complete example of a spherical astrolabe is known to exist. The astrolabe was introduced to Europe through Spain by the Islamic peoples who occupied the north coast of Africa and part of Spain in the 10th century.

The astrolabe is thought to have had its origin in the Greek communities of philosophers and astronomers shortly before the beginning of the Christian era, and it continued to be made in Western Europe until the 17th century, and until the 20th century in some Islamic countries. This was because an astrolabe had traditionally been used to determine times of prayer. The earliest surviving astrolabes are Islamic, from Syrian workshops of the 9th century. In the 10th century, Persian astrolabes were being made in Isfahan, which became an important centre for the craft. The Muslim conquest of Spain introduced the knowledge of astronomical and other scientific matters to that country. In the 11th century the craft of the astrolabe maker became recognised, and when the Christians recaptured Toledo in 1085, the Europeans developed the manufacture of astrolabes and other instruments. Centres arose in France, Germany, the Netherlands and England. The *Treatise on the Astrolabe*, written in 1391 by Geoffrey Chaucer, the English literary figure, is not only an excellent introduction to the use of the instrument, having been written for Chaucer's son Lewis, aged 10, but is also important as the first technical treatise written in English (**1**).

Around the outer edge (called the limb) of an astrolabe is a scale divided into 360°, and there is an alidade, or rule, with a pair of sighting vanes. A prominent star may be sighted and its altitude measured. A fretted disc (the rete), containing a number of star pointers, can then be turned until the measured star's pointer cuts or crosses the altitude circle on the stereographic projection (**2a**). In this way, the pattern of the stars at a particular time is set on the model. If the day of the month is known, it is possible to read off the time. The instrument is, in fact, an analogue computer. One must remember that the stars appear to rotate once in 24 hours. It is also possible to find the time of rising or setting of a given star, or of the Sun, at any date required. Another use is in surveying (Chapter 3), and a form of astrolabe was used by seamen.

European astrolabes are now rare and expensive. Genuine Islamic astrolabes are sold by the main auction houses, and within the means of a modest collector (**3–4**). One should beware, however, of indiscriminate purchase because of the large numbers that have been made this century, or even earlier, for the tourist trade. These are crude and invariably incorrect in the mathematics – the 'projection' is a collection of circles rather than truly stereographic. Since the inscriptions in Arabic are difficult to read unless you are an expert, take advice before you purchase.

Sand-glass

Also called the hour-glass, although it may tell half or quarter hours, the sand-glass is thought to have been invented in the Mediterranean area in the 12th century. It may, indeed, be associated with the magnetic compass, which was used by seamen to plot magnetic bearings and distance on a portolan chart. Certainly, it has become popularly associated with navigation, 30 minutes being an interval timed on a course and noted on the traverse board. A half-minute sand-glass was also employed with a log-line for estimating a ship's speed.

Other activities, too, would have required a sand-glass, since monastic disputations, church sermons, school lessons and law-court cases all required the facility for simple timing. Even today, small sand-glasses may be used for timing eggs.

The sand-glass consists of two pear-shaped glass flasks, joined at the necks, and part filled with dry, uniform-grained sand. The frame that holds the glass has to be reversible, and is usually of wood, very simply, even crudely, made. Other frames may be in elaborately carved wood, ivory, or embossed and gilded metal. The flasks may be individually mount-ed for 60, 45, 30 or 15 minutes, or they may be united in a row of four.

The two separate flasks were formerly joined at their necks with a perforated diaphragm between them, and then a seal was made with wax or putty, and bound with cord. This practice was changed in around 1725, when a perforated brass ball was put between the necks and the seal made by fus-ing the glass. By the end of the 18th century the double flask was blown as a single unit. The actual size is no indication of the time the sand will run, which is solely controlled by the size of the hole between the flasks and the grain size of the sand.

Unless the decoration is elaborate, it is rarely possible to date or even give a country of origin to a sand-glass, because there is scarcely ever a maker's mark, and glass and plain wood are virtually undatable. This makes modern replicas, of which there are many, easy to pass off as originals.

Quadrant

The quadrant is named from its shape, a quar-ter of a circle (**18**). The curved edge is divided from 0° to 90°, and at the right-angle, the apex, a cord is attached with a small weight of lead or brass at the end. A pair of metal pin-hole sights is mounted on one straight edge. By holding the quadrant vertically, and aligning the sights on the Sun or a star, the angle of elevation is read off the degree scale by the position of the cord, kept in a vertical line by the weight. This instrument is used for navi-gation and surveying, and the plain altitude quadrant with degree scale and plumb-line by artillery officers, for setting the angle of a gun barrel. The basic use was for time-telling.

The horary quadrant is medieval in origin. The now exceedingly rare *quadrans vetus* (old quadrant) was of brass, with a sliding, curved plate (cursor) that moved in a slot above the degree arc. Engraved on this cursor were a Zodiac scale and a solar declination scale. Above this the rest of the quadrant had a dia-gram of planetary hour (unequal hour) lines. The cursor was set to the latitude of the observer, the plumb-line was held over the date on the declination scale, and a small bead on the plumb-line was moved till it cut the 6 o'clock line (midday). With the plumb-line free, the sights were pointed at the Sun so that the shadow of the foresight fell right over the backsight, when the position of the bead told the time in unequal hours. This is the mode of action of all horary quadrants. Later they ceased to be universal and had fixed scales for a particular latitude, and hour lines for Italian or Babylonian hours. A shadow square for trigonometrical surveying was often added.

Two sand-glasses. Left: oak frame, red sand, running for one hour; late 18th century. Right: mahogany frame, white sand, running for 18 minutes; the two bulbs are blown as one piece; early 19th century. (MHSO)

Gunter's quadrant, or Gunter's astrolabe quadrant, was first described by Edmund Gunter, Professor of Astronomy at Gresham College in London, in his book *De Sectore et Radio*, published in 1623. A stereographic projection of the Equator (or the equinoctial line), the tropics, the ecliptic and the horizon, in the manner of an astrolabe projection, is engraved on brass or on a copper printing plate for paper versions. The arc is divided into 90° as before, and on the left-hand edge is a solar declination scale, 0°–23½°, which is the angular distance the Sun moves from the Equinox (when the Sun crosses the Equator) to the Solstice (midsummer or midwinter). It is, in fact, the angle at which the axis of the

Earth is tilted in relation to the orbit round the Sun. The Sun's meridian (noon) altitude on a given date depends also on the latitude, and so the meridian altitude at midsummer at a latitude of 52° North, is the co-latitude (90° minus 52° = 38°) plus the declination, 23½°, giving 61½°.

At midwinter, the meridian altitude of the Sun is 14½° (38° minus 23½°). The arc running through zero on the declination scale represents the Equator or equinoctial, and the arc at 23½° represents either of the tropics. The dotted line that starts from the zero and reaches the bottom right-hand corner of the instrument is the ecliptic, i.e. the Earth's orbit or the course through which the Sun appears to travel during the year. The signs of the Zodiac are marked on it. The other dotted line that goes from the zero to nearly half-way along the bottom arc represents the horizon.

On the left of centre are two sets of hour lines (here in modern, equal hours) that converge on the equinoctial line; those curving to the left are for the winter half of the year, and those curving right are for the summer half. On the right of centre are lines of the Sun's azimuth, where 0° is the meridian, and 90° is the East–West line. By measuring the Sun's altitude, and setting the time bead, the azimuth (horizon angle) of the Sun can be found by the position of the bead among the curves when the plumb-line is positioned over the complementary angle to the altitude on the degree scale. Again there are two sets of curves; those going left are for summer, and those going right are for winter use.

To tell the time by the Sun, the bead is positioned on the line according to the declination. This can be taken from the scale to the left or, more accurately, from the position of the Sun in the ecliptic. The altitude of the Sun is then measured, using the sights. For example, on 20 April the Sun enters the sign of Taurus (before 1582 in most Continental countries and before 1752 in Britain, the date would have been 10 April by the Julian calendar), giving a declination of 11° 14´; and if the altitude is 36°, the plumb-line is placed over this angle when the bead will cut the hour line marked 9 and 3. To determine whether the time is before or after noon, it is necessary to discover whether the Sun is still rising (before noon), or starting to set (after noon).

Time at night can be found by taking the altitude of one of the stars whose correct ascension is marked on the instrument and making a simple calculation. The month scale used with the degree scale gives the Sun's meridian altitude on any day, so, conversely, the date can be found by measuring the meridian altitude. Other calculations can also be made with this ingenious instrument.

To find the latitude for which the quadrant was designed, supposing that it is not marked, a line is taken from the apex through the 12 o'clock point and extended to the degree scale, where the co-latitude is found. Thus for a measured angle of 38°, the latitude is 52° (90° minus 38°).

There were other mathematicians and surveyors who devised quadrants, but the Gunter's quadrant is frequently encountered by collectors. Most date from 1650 to 1750. Being pocket-sized, they were popular as a convenient time-teller and almanac. In addition, they were used for teaching purposes in universities. Sizes are from about 4 inches (10 cm) to 6 and 9 inches (15 and 23 cm), and examples can be found in boxwood, ivory and often in brass. There are printed versions stuck on to oak boards. The back of a Gunter's quadrant may be blank or variously designed, either with a map of the prominent constellations, in the form of a nocturnal on a rotating disc, or as a sundial with a slot into which fits a style.

Also to be found are Islamic quadrants that have an astrolabe quadrant on one side, and on the other a sinical quadrant with arcs of sines and cosines.

Nocturnal

The nocturnal *(nocturlabe* in French) is a simple instrument for giving a rough indication of the time, perhaps to a quarter of an hour, during the night (**5b, 10**). Its use depends on being able to see the Pole Star and the Great Bear (Ursa Major) or Plough, and on the fact that the stars appear to rotate about the Pole once in 24 hours (less four minutes each day). In appearance, it looks like a small, circular table-tennis bat with a calendar scale engraved on it. Rotating over this is a disc marked with two periods of twelve hours (sometimes with night hours only, 8 pm to 8 am). The hour positions are usually notched for counting by feel, the 12 o'clock having a larger notch. Above this disc is a long rotating pointer, and the central rivet has a hole. To tell the time, the 12 o'clock is set to the date, the instrument is held upright, the Pole Star sighted through the hole, and the pointer turned to be in line with the Guards of the Bear, i.e. the two prominent stars *a* and *b* in the constellation that align with the Pole Star. The time is shown by the pointer cutting the hour disc.

The nocturnal was described in some 16th-century texts, and examples dating from about 1500 exist. Italian and French models purporting to be late 16th and early 17th century are likely to be modern replicas. Dutch and English models of the late 17th and 18th centuries, in either brass or boxwood, are generally authentic (**10**). Few are signed. Sometimes there is provision for using the star b in the Lesser Bear (Ursa Minor) as well, and stamped marks such as 'GB' and 'LB', or 'Both Bears' can be found. The earliest known date on an English wood nocturnal is 1637.

Sundials

Of all mathematical instruments, the one most commonly encountered is the sundial. The chief reason for this is that the mechanical clock absolutely required a sundial to check that it was keeping correct time. It is probably true to say that the great numbers of sundials made in the 17th and 18th centuries matched the equally large numbers of clocks in general use. Every church required a sundial to ensure that the turret clock kept time. Similarly, in the late 17th and throughout the 18th century, most country houses had a horizontal dial, usually mounted on a pedestal in the garden. Anyone who possessed a watch would very likely also own a pocket sundial, and on occasion, the two were combined in the same case. Every locality has its own time, and consequently the local time at Oxford, for example, is five minutes slower than Greenwich, because Oxford is 1¼° West of Greenwich in longitude. At Bristol, the local time is 10 minutes behind Greenwich.

Two developments occurred in the 1830s

that independently transformed the situation. The railways needed to keep the same time along all routes, and this could conveniently be done by means of the electric telegraph, which was first used in 1837. By the mid-19th century, the National Telegraph Company had offices in most towns, and exhibited a clock that told Greenwich time. From then on, the sundial ceased to be of any practical importance, and remained merely as a decorative object.

Most sundials work by casting a shadow on to a marked-out surface; the edge of a piece of metal, a rod or a string may be used to cast the shadow. Some work by letting the Sun's rays pass through a small hole, so that the time is read by a spot of light. The object that casts the shadow is called the gnomon (from the Greek word meaning 'indicator'), or the style. In general, the gnomon is arranged to be parallel to the axis of the Earth. The plate that receives the shadow can be parallel to the horizon, or vertical to this plane, or in any position, provided that it is engraved specifically for that orientation. But, on some dials the shadow is received on a ring, which is parallel to the Earth's Equator. An example of this is the equinoctial ring dial, where the hours are easy to mark out, because they are at 15° intervals round the ring.

Looking at a large collection of sundials in a museum one cannot help but be amazed by the variety of protrusions and apertures that have been devised for use in time-telling. One amusing example allows the heel of a Dutch wooden clog to cast a shadow along the sole. Although the principle of a sundial is so simple, there are many types still to be found today, each with its own peculiar characteris-

tic and descriptive name (**5a**). The most common of these types are described below.

Horizontal Dial

This is the garden sundial, which can also be found in small versions (see Butterfield dial). The hour lines are engraved on a round or square plate of brass, bronze or slate, and in the middle is the gnomon of thin metal, the straight edge of which is parallel with the axis of the Earth when the dial is set in its position. The hour lines have to be calculated for a particular latitude which is usually engraved on the South side. The latitude angle can be found by measuring the angle between the gnomon and the plate. A postcard or something similar can be placed behind the gnomon, and a pencil run along it; a protractor can then be set over the angle (see Table I).

When, from the 18th century, clocks became far more accurate, the seasonal variations in the Sun's times became extremely important because the clock keeps mean time whereas the sundial does not. This is because the Earth has an elliptical, not a circular, orbit. To allow for this, the equation of time is indicated against the date on a ring round the dial plate. Periods are marked thus: Watch Slow and Watch Fast; the variation can be as great as 18 minutes.

Good quality garden dials were made by established mathematical instrument makers, such as Elias Allen, John Marke, Edmund Culpeper, Benjamin Scott, Benjamin Martin and Peter Dollond. There are many modern horizontal garden dials, some of which are purely ornamental. Simulations of old dials were made, particularly in the 1920s, that seem at first glance to be genuine, and some bear a year in the 17th century. Beware of any dial with a Sun's face, and a tag such as 'I tell ye sunny hours', or 'Set me right and treat me well, And I the time to you will tell': these assuredly are modern.

Vertical Dial

This is the kind very frequently found on the wall of a church, probably incised into the stone of the structure. Other examples may be of painted wood, brass, sandstone or slate. Those that were made to face due South may resemble horizontal dials if removed from the wall. But a vertical dial cannot receive the Sun's light before 6 am or after 6 pm at any time of year, so the missing hour lines reveal its type. Small versions are found on diptych dials. If the wall is not due South, the gnomon is skewed, and the 12 o'clock line is at an angle to the edge of the plate.

Polyhedral Dial

This type can combine horizontal with vertical in several orientations, and may include plates at an angle with the vertical. Twelve-sided solids can have each face fitted up with a gnomon, and may be found as monumental dials in grand settings. Small brass and even pottery versions are known, a popular late 18th century type being a cube. This was produced in Germany by David Beringer of Nuremberg, among others, and had coloured printed paper dial faces glued on to the smooth wooden cube.

Compass Dial

A horizontal dial over a magnetic compass, this type goes back to the 15th century. The dial plate has been cut away to reveal the compass card and needle, and the gnomon is hinged for packing. Andreas Vogler of Augsburg manufactured them in the 18th century. The popular compass dial was quite small, and is still made as a novelty.

Magnetic Dial

In name, this is easy to confuse with the compass dial, but in the case of the magnetic dial, the plate with its gnomon is the actual compass card. Below the pivoted card, and fixed to it, is the magnetic needle. This means that the dial is self-orientating; at least, it will orientate to the magnetic North, and a correction of some sort will be necessary for declinations. These small dials became quite popular in the first half of the 19th century, being produced by the Fraser family and Samuel Porter, both in London.

Butterfield Dial

A form of horizontal dial with an adjustable angle to the gnomon and a compass, this semi-universal type is named after Michael Butterfield, who worked in Paris at the end of the 17th century. This very popular pocket dial was frequently made in silver, but also in brass. The characteristic shape has an eight-sided plate, with one pair of sides longer than the others; an oval version is also found. French-made models have four hour rings, one inside the other, for latitudes 43°, 46°, 49° and 52°. The gnomon can be raised to suit these latitudes, the degree pointer being the beak of a bird. English-made Butterfield-type dials have only one hour scale for 52°, but the bird pointer is copied. A small compass is incorporated in the plate at the South end. The plate is supported by the base of the compass box and two feet. On the underside are lists of towns and their latitudes. Many were made during the 18th century, and signed by various craftsmen in Paris and London. Butterfield's name was forged in the 18th century, and this may also occur today.

Names found on these dials include Butterfield à Paris; P. Le Maire à Paris; Cadot; Le Maire Fils; Macquart; J. Simons, London; E. Culpeper, London; Richard Whitehead, London (*c.* 1690); Nicolas Bion, Paris; and Charles Bloud, Dieppe.

Inclining Dial

This is a portable universal dial, based on the compass dial. Some are small enough for the pocket, others are suitable to be placed on a window-ledge. A large compass is in the base, for quick orientation. The hour plate is hinged, and can be set at an angle read from a curved arm fitted to one side of the base. This dial is handy for travellers, and the underside of the base usually has a list of principal towns with their latitudes. The gnomon and the arm are hinged for packing flat. The principle is that any horizontal dial can be used at another latitude provided the shadow-casting edge of the gnomon is parallel to the Earth's axis. English examples from the early 18th century were made by Bryan Scott, Jonathan Sisson, Heath & Wing and Dollond. Others include Johann Martin (*c.* 1700), Augsburg; Pierre Le Maire (*c.* 1750), Paris; and Francis Morgan (*c.* 1780), St Petersburg.

Analemmatic Dial

This is the name given to a special form of horizontal dial, so made as to be self-orientating. The word 'analemma' signifies the table of the Sun's daily declination from tropic to tropic (23½°N to 23½°S) and back during the year. In addition to a horizontal dial for a given latitude, there is another dial with a vertical gnomon that is attached to a slider which can set the gnomon to any date in the year – this is, in fact, the analemma. Around this is an hour scale in the shape of an ellipse, and when the sundial is positioned so that both hour scales tell the *same* time, then it is orientated to the meridian, and the time is known. Pierre Sevin of Paris made such a dial in about 1670, and Thomas Tuttell of Charing Cross, London, introduced it to Britain, describing it in a little book published in 1698. Although it is a rare type, a fine specimen exists by John Bird, the famous 18th-century astronomical instrument maker, and Victorian examples are also known.

Diptych Dial

The word means anything folded so as to comprise two leaves (**6**). Here two plates are hinged together, and in use are opened out to make a right angle: a string gnomon is thus drawn taut and the shadow falls on to a horizontal dial as well as a vertical dial. These are cut on the inner faces of the plates. There is usually a magnetic compass, and pin gnomon dials for Italian and Babylonian hours. On the outside of the top plate may be a wind rose and pointer, and on the underside of the bottom plate is customarily a lunar disc with calendrical information. The peak output for this type was at Nuremberg in the late 16th and early 17th centuries, and the traditional material for their construction was ivory (**8**). They vary in size and may be elaborately decorated, the cuts being filled in with a black, blue, red or green colouring agent. Although not rare, they are ancient and attractive, commanding high prices today. Famous makers include Georg Hartmann, Hans Tucher, Hans Troschel, Jacob Karner, who used as a mark the figure 3, and Lienhart Miller.

Late 18th- and early 19th-century German versions in wood with printed scales were also produced in large numbers, some showing the latitudes of principal towns in the United States of America.

Magnetic Azimuth, or Bloud Dial

This is a form of diptych dial, nearly always in ivory, which was produced by Charles Bloud and others working in Dieppe, chiefly during the period 1650–70. There may be a string gnomon dial, but generally the top plate is a polar dial, whereby the plate is propped up according to the latitude so that it is parallel to the Equator, and a rod is inserted at the centre of the hour circle.

The main feature is the magnetic azimuth dial, so called because when the dial is turned to the position where the shadow of the top plate falls exactly over the lower, the magnetic needle points to the time on an elliptical, pewter hour scale. To allow for the declination of the Sun, this hour scale has to move in a slot, operated by turning a metal disc on the underside of the bottom plate to the appropriate date. The disc is always engraved with a perpetual calendar.

This type of dial works effectively only when the magnetic declination is zero, which was the case for Dieppe, London and neighbouring areas in 1657.

Names of other Dieppe craftsmen include Jacques Senecal and Ephraim Senecal.

Universal Equatorial, Equinoctial or Augsburg Dial

The hour scale is designed to be parallel to the Equator, and when the Sun is exactly over the Equator it indicates the time of the Vernal or the Autumnal Equinox, hence the basic names given to this dial (**7a, b, 9, 14**). Being universal (i.e. adjustable for latitude) it was a popular type made in large numbers, especially in Augsburg, Germany, during the late 17th and 18th centuries. It is usually made of brass, and in its German form is often eight-sided. In the base is a magnetic compass, and on the South side there is generally a levelling bob in a hinged frame.

On the West side is a curved arm engraved with degrees of latitude. The hour scale is cut on the inside of a thin ring, and since it is equatorial, the divisions are all equal, at 15° to one hour. The gnomon is a needle on a cross bar which can be turned so that the needle is at right angles to the ring. Everything hinges for packing into a fishskin case to go in the pocket. A perpetual calendar can be an addition to the lid of the case.

Some of the 18th-century Augsburg craftsmen who made this type of dial are Nicolaus Rugendas, Johann Georg Vogler, Andreas Vogler and Ludwig Theodatus Muller. Two 18th-century Parisian makers are Claude Langlois and Macquart & Cadot.

Crescent Dial

This is a variation on the equinoctial dial: the hour circle is divided in two and the parts are transposed so they virtually touch at the 6 o'clock points, making a double crescent (**15**). The gnomon is also in the form of a crescent, with the tips casting the shadow. The crescent is held at the middle by a screw, which can move in a slot with a declination scale. This adjustment for the time of year means that the shadow of the tip of the gnomon falls exactly on the hour ring; so the instrument is turned until this occurs, making the dial self-orientating, which obviates the need for a compass. There is a levelling bob or spirit level and adjustment screws in the base plate. A decorative dial, it can be made in brass, gilded brass or silver. It originated from Augsburg in the late 17th century. Notable makers are Michael Bergauer, Johann Martin, Nicolaus Rugendas and Johann Willebrand. The signature 'P. Masig à London' means a Martin dial sold by his agent. French models are known, but the crescent was never popular in Britain.

Universal Mechanical Equinoctial Dial
This form of dial, similar to an Augsburg dial, is intended to read to a minute of time (**16**). A normal dial, approximately 2½ inches (6 cm), can read to about a quarter of an hour, but by providing a geared mechanism, a small movement of the alidade is magnified so that time can be read more accurately on a separate minute dial. The earliest was produced in 1671 at Augsburg by Michael Bergauer.

Universal Equinoctial Ring Dial
Self-orientating and universal, this is one of the most elegant and accurate of all sundials (**11**). Since the hour ring is in the plane of the Equator, the accuracy of the scale is high because the divisions are equal, and with large models of say one foot (30 cm) in diameter, made by the best craftsmen, time can be read to about one minute. There are three principal parts: the outer meridian ring; the inner hour ring; and a central bar. The meridian ring has a suspension ring adjustable for latitude. The hour ring is pivoted at the 12 o'clock positions to the outer and is set at right angles to it. It is divided into 24 hours on the inner side. The bar which is pivoted to brackets fixed to the outer ring, has a slot with a declination scale, months on one side and the signs of the Zodiac on the other. In the slot is an index with a pin-hole. Turning the instrument causes a spot of light from this hole to fall on the hour scale. When the spot is exactly on the scale, the time is given and the outer ring is positioned in the plane of the meridian, i.e. it lies North-South. The back may have a solar altitude scale 0°–90°, used with a pin pushed into a hole in the outer ring. This type was invented after 1600, possibly by the English mathematician, William Oughtred. The names of just about every mathematical instrument maker appear on examples of this type of dial, which was made all over Europe.

Larger versions are known, mounted on an azimuth plate with a pair of spirit levels and three adjusting screws, and a magnetic compass. Thomas Heath in the mid-18th century preferred this model (**12**).

Ring or Poke Dial
This is the simplest of dials. However, the accuracy is poor. Made as a wide ring of brass, it has a sliding collar with a pin-hole to let a spot of light fall on the graduated inner side of the ring when suspended vertically. The collar adjusts to the solar declination. Simpler versions have a fixed hole and two scales for winter and summer; others have a hole on each side of the suspension and two seasonal scales opposite them. A cheap dial, it was popular

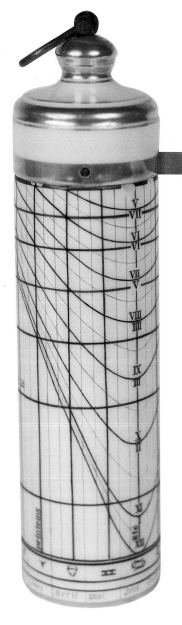

An unusually large, mid-19th century, French pillar dial, height 9¾ inches (24.7 cm), made in white porcelain with gilded rims. The hour lines are drawn for the latitude of 49° 35′. When the horizontal gnomon is swivelled over the date line and pointed to the Sun, the shadow tells the time. (WMHS)

during the 17th and 18th centuries among country people who kept it in their poke, or pocket. French, Italian, German and English versions exist, often unsigned. Luke Proctor of Sheffield did a good trade in these dials in around 1800. Other English ones are signed by initials only: E.E.; I.H.; T.W.

Pillar, Cylinder or Shepherd's Dial
The hour scale is marked on the outside of a cylinder which stands vertically. The gnomon projects horizontally from the top, and when not in use is kept inside the pillar. The depth of the scale varies from winter to summer, and the gnomon has to be set over a month or Zodiac scale. A pillar dial may be very simple indeed, in plain boxwood with carved lines, a type in use in the Pyrenees until the 20th century. Others may be of wood with printed scales, and a few are in ivory, made in Dieppe during the 18th century.

Scafe, Cup-dial
This is one of the oldest known forms of sundial, Roman examples having been preserved. Here the hour lines are engraved on the inner surface of a hemisphere or even a metal goblet. The gnomon is generally vertical, but sometimes parallel to the polar axis. The tip tells the time. Shallow scafes are to be found on some ivory diptych dials. Italian examples from the 16th century are known, in particular by Alexander Ravillius in ivory, dated 1537, and Hieronimus Vulparia, 1588. Georg Hartmann of Nuremberg made one in brass, dated 1539.

Cannon Dial

This is a horizontal dial, usually on marble or stone, with a miniature cannon and two brackets that support a burning lens. The bracket arms are set to the solar declination, so that when it is noon the burning beam of light fires the gun. Some large specimens are for use in ports or camps – the noonday gun – but most examples measure only 3–9 inches (7–21 cm). The cannon dial was patented in the early 1800s by Victor Chevalier, an instrument maker of Paris. In 1880 they were offered for sale at 3 guineas (£3.15) by Negretti & Zambra of London. A large model is signed by F. Amuel of Berlin, and another German one is no bigger than a pocket watch.

Dipleidoscope

In March 1843, Edward John Dent, the noted chronometer maker, put on the market a newly patented device for noting the meridian passage of the Sun with great accuracy, which he named the dipleidoscope ('double image viewer'). The invention was by James Mackenzie Bloxam, who patented it as a 'meridian instrument' on 20 June 1843 (No. 9793). It consists of a hollow right-angled prism, with two sides silvered and one of glass. The meridian transit was known by the coincidence of two images of the Sun by single and double reflection, one from the top glass and the other from both mirrors. Once the base of the instrument was accurately levelled and orientated, the time could be read to seconds. The instrument was made in fixed and portable versions; the latter had levelling screws and bubble levels, latitude adjustment and a magnetic compass needle. Dent cannot

have profited greatly from his novelty because the new electric telegraph soon replaced the need for an accurate time check in the form of a sundial. Nevertheless, it remained important to explorers in remote parts of the world.

Armillary Sphere

This instrument, whose name derives from the Latin word for 'bracelet', consists of a series of concentric rings representing the great circles of the heavens (**17**). It was to be found in early observatories, equipped with movable sights, and was used to measure the position of a celestial body with reference to co-ordinates on either the equatorial or the celestial circle. The armillary is best known, however, in its more convenient portable form as a teaching and demonstration instrument, examples of which survive from the 15th century. It was either hand-held, with the rings mounted on a polar axis that extended into a handle, or mounted like a globe in a horizon ring supported by a stand.

The early armillaries were based on the Ptolemaic concept of the universe. In the middle was a ball to represent the Earth, surrounded by rings for the Poles, the Equator, the tropics of Cancer and Capricorn and the ecliptic band, generally engraved with the signs of the Zodiac. The Greeks believed that the Sun, Moon and planets revolved round the Earth, against the background of the fixed stars. Then, in the 16th century, Nicolas Copernicus described the true nature of the solar system, with the Sun at the middle. This was a controversial area, but as it became accepted, two armillaries were used to demonstrate the old and new solar systems.

The sphere was used by both teachers and lecturers, and during the 18th century often found a place in libraries as a companion to globes. It varied in size from the large, floor-standing models to such delicate examples as one made in silver and ebony by John Rowley of London that is only 6 inches (15 cm) in diameter. Although generally made in brass, with a wooden stand, armillary spheres were also produced in pasteboard and wood in the late 18th and early 19th centuries – a cheap alternative that made this popular instrument accessible for use in the schoolroom and by the amateur astronomer.

Because of its elegant and unmistakable appearance, the armillary sphere became a symbol of the astronomer, appearing on the title-pages of astronomical manuscripts and books, and used by painters as appropriate furniture for a scholar's study. Prince Henry the Navigator of Portugal chose it as his emblem in the early 15th century, as recognition of the importance of astronomy to the deep-sea sailor.

Planetarium

An instrument that shows the motions of the bodies in the solar system is called a planetarium. It is a descendant of the armillary sphere. Small ivory balls representing the planets are supported on brass wires, with long rods to a

Two examples of the Dent dipleidoscope. *Left:* signed: E.I. DENT LONDON PATENTEE; *c.* 1870. There are cross bubble levels and a magnetic compass, with a latitude scale. This model can give an accurate time check to an explorer. *Right:* signed: *E.J. DENT PATENTEE 752; c.* 1845. The prism is in a fixed mount for one latitude, stamped: INDIA. (MHSO)

collar round a central pivot. The rods have to be pushed round by hand in the case of the wooden models, and by a crank and gearing with the brass 'drum' type, devised by Benjamin Martin in 1747.

Some are very large, 4 feet (1.2 m) or more in diameter, and feature elaborate clockwork mechanisms, craftsman-made cases and glazed covers. Other planetaria are often termed orreries, so named after Charles Boyle, 4th Earl of Orrery, who had a planetary model made for him by John Rowley in 1712. Rowley had been influenced by a clockwork device constructed by Thomas Tompion and George Graham in about 1709, which had a small globe of the Earth and an ivory Moon, to show both the daily and annual motions round the Sun; strictly, this is a tellurium, but the naming of these astronomical models is confused, and orrery has become the generic term.

In Britain, the glazed-cover planetarium mentioned above is known as a grand orrery. Planetaria were also sold with alternative systems to be placed on the central axis; a large brass ball for the Sun was always superimposed. The planetary system was removed, to be replaced by a 'tellurium', the Sun–Earth system, or a 'lunarium', the Sun–Earth–Moon system, stressing the Moon's orbit. There are hybrid instruments with combinations of all three variations permanently arranged.

A small orrery, c. 1810, made of printed paper mounted on oak, contained in a box covered in marbled paper. The instrument demonstrates the motions of the Earth and Moon relative to each other and to the Sun. Such demonstration models were produced by W. & S. Jones of London. (MHSO)

George Adams of London exported to Holland in 1790 a 'drum'-type planetarium which showed Mercury, Venus, Earth and Moon, Mars and Jupiter with four moons, Saturn with ring and eight moons, and Uranus with two moons. The number of planets and of moons helps to establish dating around the end of the 18th century. Herschel discovered Uranus in 1781, and its two moons in 1787. Saturn was found to have a seventh moon in 1789, but its eighth moon was not discovered till 1848 (Adams had anticipated that Herschel would discover the eighth, and made his instrument accordingly). Neptune was discovered in 1846.

A knowledge of elementary astronomy was in great demand during the early 19th century, and many planetaria exist from this time. The usual form is a round wooden disc, covered with a printed and coloured sheet, showing the calendar and Zodiac, and giving information about distances from the Sun and the occurrence of comets. Sizes vary, some items being no more than toys.

Names associated with this type include W. & S. Jones, William Cary, William Harris and Newton & Sons.

Globes

The history of charts, maps and globes is a specialist study which, although connected with navigation and surveying, cannot be dealt with fully in this book. What is important to know here is that large globes of 2-feet (60-cm) diameter or more were used by navigators to help solve problems in sailing. Pairs of globes, one of the Earth and one of the constellations – terrestrial and celestial – were regular furnishings in a library from the 17th century on. During the 18th and 19th centuries they had a place in nearly every school, and common sizes were 1, 1½ and 2 feet (30, 45 and 60 cm).

Small pocket globes, from 2–3 inches (5–8 cm) in diameter, are available to collectors. They are made from pasteboard or wood, covered with gores printed from engraved copper plates, and hand-coloured. At the poles are brass pins that support the globe in its case. This is composed of two hemispheres hinged together, which are covered on the outside with black fishskin, and on the inside with the celestial sphere, showing the constellations.

The pocket globe was first produced by the London mathematician and hydrographer, Joseph Moxon, who also sold instruments, maps and globes. He produced, among others, 9-inch (25-cm) pocket globes before 1700. In the 18th century the trade in these small globes was continued by Charles Price, John Senex, Robert Cushee, James Ferguson, Nathaniel Hill, and George Adams.

During the early 19th century, important makers and retailers are Dudley Adams, William Cary, John Newton, Charles Schmalcalder and James Cox. The manufacturers of these globes were fewer than the names found on them.

The pocket globe was mainly used for teaching geography and astronomy to children; but there was probably a subsidiary use as an *aide mémoire* to an educated person who could not afford a large globe. Pocket globes enjoyed a period of high production at the time of the voyages of discovery, for example, by Captain Cook to Australia and New Zealand, and George Vancouver to the west coast of America.

The small globes may have served as references and conversation pieces in both social and geographic circles as each new map edition charted a new track of Cook's, or a recently defined far-off coastline.

It is in this way that the globes can be dated, even if the engraving of the date of the original production has not been altered. George Anson sailed around the world in 1744, Cook's three voyages occurred between 1768 and 1774, and Vancouver explored western Canada in 1790–5. The other factor in dating is the development of the coasts of Australia and North America. The island of Tasmania was shown connected to the mainland until 1792. On a Hill globe of 1754, the coast of Alaska is missing, and Greenland is joined to eastern North America.

TABLE 1

Horizontal Sundial

Angles between hour lines (A) at nine latitudes

Hour	Hour angle	Latitude								
		44°	46°	48°	50°	52°	54°	56°	58°	60°
12	(0°)	0	0	0	0	0	0	0	0	0
1	(15°)	10.5	10.9	11.3	11.6	12.0	12.3	12.55	12.8	13.0
2	(30°)	21.85	22.6	23.2	23.85	24.5	25.0	25.6	26.1	26.6
3	(45°)	34.8	35.7	36.6	37.5	38.2	39.0	39.7	40.3	40.8
4	(60°)	50.3	51.25	52.2	53.0	53.8	54.5	55.1	55.75	56.3
5	(75°)	68.9	69.6	70.2	70.7	71.2	71.7	72.1	72.5	72.8
6	(90°)	90.0	90.0	90.0	90.0	90.0	90.0	90.0	90.0	90.0
7	(105°)	111.1	110.4	109.8	109.3	108.8	108.3	107.9	107.5	107.2
8	(120°)	129.7	128.75	127.8	127.0	126.2	125.5	124.9	124.25	123.7
9	(135°)	145.2	144.3	143.4	142.5	141.8	141.0	140.3	139.7	139.2

For angle between hour lines A, latitude L, hour angle H: $A = \mathrm{Tan}^{-1}(\sin L \times \tan H)$

NAVIGATIONAL INSTRUMENTS

ORTUGUESE voyages of discovery to Africa and, later, the Far East first encouraged serious interest in positional navigation. Mathematicians, astronomers and cartographers were brought to the training workshop for officers, established in 1438 at Sagres, in Portugal, by Prince Henry the Navigator, to teach the science of navigation, to devise instruments and to draw up charts. Following the discovery of the American continent by Christopher Columbus in 1492, the skills and the instruments needed for successful ocean sailing became of practical and economic importance.

While a ship is sailing along the coast, the determination of its position is not a great problem. Beginning at its port of departure, which has a known longitude and latitude, the course and the distance can be plotted on the sea-chart. The course is steered, and the coastline and marker points, headlands, churches, rocks, are each noted. After a time, the ship's new position can be checked by taking bearings in relation to the shore. A ship's passage through the water differs from the distance travelled over the sea bottom because of currents and tidal flow, and drift caused by the wind.

In the open sea, out of sight of land, a different type of navigation has to be applied. The latitude (the angular distance from the Equator 0° to the Pole 90°) can be measured by a quadrant or cross-staff, taking the elevation above the horizon of the Pole Star. The altitude of the Sun at noon, when it crosses the meridian, can be taken by a quadrant or sea astrolabe, cross-staff or octant. Knowing the day of the month, the altitude, and hence the latitude, can be found from tables. The longitude is difficult: this is the angle around the globe, and nowadays the prime meridian is taken as the longitude of Greenwich, 0°; New York is 74° West and Hamburg is 10° East. Longitude can be measured by a clock, because if you know the time at your port of departure, and can find out the local time, say noon, by the Sun, then the difference in time gives you the longitude, because one hour equals 15° of longitude.

Clockwork mechanisms, however, were not good enough to withstand the ship's motion and changes in temperature, until after John Harrison perfected the marine chronometer in the mid-18th century. Before this, seamen had to rely on the compass, the log-line and the traverse board to keep a tally.

Charts and globes, with scales and dividers, were used by navigators to help solve the problems, which needed more skills and hence training as the centuries rolled by.

Altitude Quadrant

In the mid-15th century, Portuguese sailors used the plain altitude quadrant for angle measurements. The astronomical quadrant is known from classical times; for ship use it was about 10 inches (25 cm) in radius, was engraved with a scale from 0° to 90°, and had a pair of pin-hole sights, and a plumb-bob, which marked the angle indicated by the sights. The Portuguese navigators knew that they could return to Lisbon by taking the winds out into the Atlantic until the latitude of Lisbon (38° 42′) was reached, as shown by the altitude of the Pole Star. Then it remained to run East 'down the latitude'. The quadrant was still used on shore till the 18th century. Near the Equator the Pole Star is too close to the northern horizon to be of use, so the altitude of the Sun was the important measure.

Mariner's Astrolabe

For plotting the altitude of the Sun near the meridian this instrument came into use around 1470. It is a development of the much earlier astronomer's planispheric astrolabe. The mariner's astrolabe has the same angular divisions on the outer ring, but is not always divided right round, and it has an alidade, or sighting rule. Here, however, the similarity ends. The sea astrolabe is heavy, especially at the bottom, and is made of brass or bronze (**19**). The weight keeps it steady, and there are

portions cut away to reduce wind resistance, which would otherwise make its use harder. The instrument is held by its ring, just above the 90° mark (sometimes 0°), and the alidade is turned until a beam of sunlight from the hole in the upper vane exactly falls into the hole in the lower. The angle of elevation of the Sun can be read from the scale on the rim.

The instrument presently in the Dundee Museum, Scotland, is the oldest surviving dated example, and is marked 1555. It has a diameter of nearly 9 inches (23 cm), and it weighs 6 lb 6 oz (2.86 kg). Sea astrolabes are found in wrecks (e.g. off Northern Ireland in 1968), or among rocks (e.g. Lyme Bay, Dorset in 1967). These must have been standard equipment on ships of the Spanish Armada. The sea astrolabe is rare; only 35 were known in 1979, but the number rose to 65 in 1987, most of the latter discovered through recent nautical archaeology.

Cross-staff

As an astronomical and then as a surveying instrument, the cross-staff may be traced from the description published in 1328 by the French mathematician and astronomer, Rabbi Levi ben Gerson. His instrument, which depends on the principle of similar triangles, was called Jacob's staff after the Biblical story in Genesis 32:10. It consists of a rectangular staff, 5–6 feet (1.5 or 1.8 m) in length, with a perpendicular vane that moves over it. The staff is graduated trigonometrically so that angles can be measured by holding the staff to the eye and moving the vane until its ends are level with the points that are to be measured. The instrument was introduced into England by John Dee in the 1550s, when it was developed to measure the angles between stars, and also the heights of buildings or the angles between topographical features. Early in the 16th century it became a seaman's navigational instrument, its use being pioneered by the Portuguese. The original instrument had one vane or cross, but seamen added other vanes because for their purposes the staff had to be shorter, about 2½ feet (75 cm) long. The seaman's staff had long, medium and short vanes, of about 15, 10 and 6 inches (37, 25 and 15 cm). The staff was calibrated directly in degrees for ease of use on board ship. The typical computation was to find the latitude by measuring the altitude of the Pole Star above the horizon. The Sun's altitude could also be determined, but this required the observer to face the Sun, with the result that its light was in his eyes. To avoid this, the back-staff was devised, and the cross-staff became known as the fore-staff.

Back-staff

This was invented by the English sea captain, John Davis in about 1594, and was intended to be an improvement on the quadrant, mariner's astrolabe and cross-staff for finding the meridian altitude of the Sun (**21**). The back-staff was also known as the Davis quadrant, and as the English quadrant by continental seamen. The name quadrant arises because 90° can be measured although there is no 90° arc. The instrument looks like a large triangle with a 30° arc at one end, and a small 60° arc at the other. One-sided vanes with pin-holes move over these arcs, and at the end opposite the large arc is a push-on vane with a slit through

which the horizon can be viewed. By adjusting the vanes, the Sun's angle can be determined. The cross-staff is rare, but the back-staff is not; 18th century examples are found in sales, and are usually signed and dated. Both types of staff have been replicated and it is not always easy to distinguish the reproductions.

Octant

The back-staff was transformed into the octant by John Hadley who published his invention in the *Philosophical Transactions of the Royal Society* in 1731 (**22**). The novelty was the use of a mirror mounted over the pivot of a radial arm that moved over a graduated arc. Viewing was through a pin-hole sight at a half-silvered mirror, which caught the reflection of the Sun from the first mirror on one half, while the horizon could be seen through the clear half, thus 'bringing the Sun down' to the horizon.

The name of this instrument sometimes causes confusion because it can measure 90°, a quadrant of a circle; the actual arc, however, occupies only one-eighth of a circle, an octant. The reason is that the use of a mirror halves the angle through which the radial arm moves, although the arc itself is calibrated from 0° to 90°.

The octant came into general use after 1750, and it continued until about 1900, and later in coastal navigation. The *Deutsche Seewarte* was certifying wooden octants until 1925. It varies little in form, but the examples vary in size from 7½ – 20 inches (19 – 50 cm). The frame is mahogany in the earlier examples, and ebony after 1800. The scale may be engraved on boxwood, ivory or brass. The early ones have transversals, or diagonal scales, to measure fractions of a degree; verniers came later. Those verniers with a central zero are usually before 1780 and those with a zero on the right are after this date. Often an octant is engraved with both the name of the maker and of the owner; occasionally there is a date as well.

Sextant

With the introduction, in 1767, of the 'lunar distance' method for finding the longitude at sea by measuring the distances of certain stars from the Moon, the greater accuracy required led to the development of the sextant for use at sea after about 1770. The name refers to the actual arc, which occupies ⅙ of a circle, and not to the angle that can be measured. As with the octant, the mirror halves the angle, and the actual arc is calibrated from 0° to 120°. Much more accurate than the octant, and therefore more expensive, sextants were used by officers of the more wealthy companies such as the East India Company. Some early examples are in ebony; mostly the construction is in brass. The greatest practitioners in precision instrument making improved the accuracy of the scales on the sextant: John Bird (d. 1776), Jesse Ramsden (d. 1800) and Edward Troughton (d. 1836).

From 1800 onwards, sextants were made in large numbers; English products were sold through agents in Denmark, in the United States and elsewhere, and there were manufacturers in Paris, Hamburg, Amsterdam and other places. Efforts were made to produce sextants that could be highly accurate wher-

ever they were used in the world, and whatever the conditions. The original brass bar construction gave way to various strutted constructions which included grids, ovals with straight bars and three circles between the two outer limbs and the arc.

Another type was the double-plated form, with two thin plates held together by a series of cylinders (pillar sextant), patented in 1788 by Troughton. On more than one occasion, Troughton was known to manufacture his sextants with scales divided on silver, gold or even platinum.

A well-known Dutch contemporary of Troughton was Gerard Hulst van Keulen of Amsterdam, who produced fine sextants, which he numbered serially, as did Ramsden, Berge, Troughton, and Troughton & Simms.

Reflecting Circle

The ultimate in accuracy, exceeding that of the sextant, was achieved by the reflecting circle. This was invented by the German astronomer, Tobias Mayer, in the 1750s in order to improve the accuracy of finding the longitude by lunar distances. The principle is the same as that used in the sextant, but the arc is taken into a full circle of 360°. The sighting telescope and the horizon mirror can be moved on an arm to any position on the circle where it is clamped, and the index arm turned to effect the apparent conjunction of the two objects being measured. The angle is the difference between the readings, and for extra accuracy verniers are fitted to the arms.

The measurements can be repeated at different parts of the circle, which again helps to reduce errors of collimation and scale division.

With the circles made by Edward Troughton from 1796, three index arms with verniers were fitted, and an average taken of the three readings.

The reflecting circle was more popular with the navies of France and Germany than with British seamen, and the higher cost of manufacture limited the market.

The instrument is also known as the Borda circle after the French inventor, Chevalier de Borda, who published a description of it in 1787. Other types were made during the 19th century, the principal manufacturers including Berge and Dollond, both of London, Gambey of Paris (**29**), Dolberg of Rostock, and the company Pistor & Martin of Berlin.

Artificial Horizon

The sextant and circle are not always used on board ship; they are sometimes used by navigators on shore, and by land surveyors and explorers. In these conditions, an artificial horizon is necessary, and this can be provided by a level flat surface.

There are two common forms of the artificial horizon. The older is the trough of wood or iron into which liquid mercury is poured from a stone bottle. The liquid metal is highly reflective, and naturally forms an absolutely flat surface. To prevent wind ruffling it, the trough is covered by a roof-like triangular box with slanting sides of plate glass. The whole set is packed into a wooden box, and is often not recognised for what it is.

The same applies to the later form (still in production), which consists of a plate of black glass supported by three levelling screws, with a bubble level attached to the plate.

Bubble Sextant

At the end of the 19th century, when ballooning had become popular, the bubble sextant was devised. This also found wide use after the aeroplane made its first extended flight in 1903. It is a sextant with a bubble level fitted to one limb, so that instead of viewing the horizon, the observer looks at the image of the bubble in a mirror placed at 45° above it. The bubble acts as the horizon marker, making it an obvious navigational benefit for pilots.

Mariner's Compass

The term 'mariner's compass' is usually taken to mean the magnetic compass, but it originally meant the division of the circle of the horizon into 32 points.

The four cardinal points of the compass are North, East, South and West, and the divisions run N., N. by E., NNE., NE. by N., NE., and so on. These are the wind directions, or rhumbs of the wind. Thus, a rhumb-line is the direction followed by a ship sailing on one course, and this can be plotted on a chart. With 32 principal points, the angular distance between two successive points is 11° 15′, and with suitably constructed charts, it was possible to sail along a rhumb-line by using a compass, and so to simplify the task of the navigator and helmsman. The compass rose (i.e. the printed compass card) adopted the wind rose.

Magnetic Compass

The Pole Star, Polaris, was the seaman's lodestar (star that shows the way), and so the magnetic stone that magnetised the needle on the new compass was called a lodestone (also loadstone). Knowledge of the direction-finding property of the lodestone came from China in the late 12th century.

Magnetic compasses, using a single needle and not a card, are found on some sundials of the 15th century. By the 16th century, the navigator's magnetic compass was a soft iron wire bent to a lozenge shape stuck to the underside of a circular card which was suspended on a pin. The lodestone was necessary to remagnetise the needle as it weakened. Dr Gowin Knight invented and, in 1766, patented his artificial, compressed powder magnet, and made steel needles for the compass.

The mariner's compass assumes a number of forms. Early ones are in a circular box of wood, then brass takes over. A hanging compass in copper or brass is one intended to be hung over the master's bunk and read from underneath.

Azimuth Compass

This may be large, consisting of a brass case mounted on gimbals which contains the rose, and a sight and string gnomon on the top of the case (**20**). The rule attached to the sight can move over a degree scale from 45°–0°–45°. This was a popular instrument in the 17th and 18th centuries, and was essential for determining the deviation of the compass from true North. The magnetic North varies considerably over the globe, and a check has to be made by finding when the Sun is due South, or by finding its bearing at sunrise, when the latitude and the date would also need to be known. Fine, large azimuth compasses were made by Walter Hayes,

Richard Glynne and J. Fowler, all of London, and by Benjamin Ayres of Amsterdam. Small examples exist, made in about 1800 by Spenser Browning & Rust, the well-known octant scale dividers.

Binnacle

The binnacle was the place near the helm where the compass was kept, and the name became attached to the cupboard or box with glass lid in which the compass was housed. Sometimes rectangular, those of the later 19th and 20th centuries are cylindrical, and can be 4 feet (1.2 m) high if deck-mounted, or shorter if mounted elsewhere. These binnacles are of mahogany or similar dense wood, with a rounded brass hood containing a glass port to view the compass card. There are also small compartments at each side to contain a lamp. The card found in many of these binnacles is probably to the design of Lord Kelvin, the famous physicist, who in 1873 developed a superior system of very light magnetic needles, grouped parallel to one another, attached to a card rim that carried the points of the compass marked on it. James White, Glasgow, was at first the only maker.

When used on iron ships, which have their own magnetisation, correctors have to be fitted to the binnacles. These are spheres of iron on each side of the case.

Another improvement that is sometimes found is the card floating in liquid; this reduced wobble in rough seas. It was introduced by E. J. Dent, the chronometer maker, in 1844. As with chronometers and all other navigator's instruments, the retailer's name, not the manufacturer's, is the one found.

Logs

Until the 16th century, seamen estimated the speed of their ship through the water by experience. Then some threw out an object and timed it passing two fixed points on board – a rough estimate of timing over a measured distance. In the mid-16th century, the 'English log', where a log or lump of wood attached to a line was thrown overboard, was produced. The length of the line run out in half a minute (measured by a sand-glass), gave the measure of the speed. By tying knots in the line at every seven fathoms (1 fathom = 6 feet [1.8 m]) a count of the knots streamed out in half a minute charted the course of the ship in miles per hour. It was then calculated that 60 miles made one degree on the meridian – a crude measure, but any standard was better than none.

From the log-line is derived the word 'knot' for the speed of a ship. Nineteenth-century log-lines on their reels, and perhaps their floats, may be found in sales, but they are not common.

Mechanical logs were proposed by several leading inventors in the 18th century, but the first successful recording log was that patented in 1802 by Edward Massey, the Newcastle instrument maker. These logs depend on rotors, and registering dials.

The 'Dutch log' uses the old-fashioned method of timing a piece of wood past a measured distance on the ship's side. The latter technique was employed by Dutch seamen from the 17th to the 19th centuries, the 'aid' taking the form of a brass tobacco box, rectangular in form, with rounded ends. There is usually a perpetual calendar and two engraved figures on the lid, and underneath are engraved the speed tables that convert the time measured into speed.

Traverse Board

The helmsman's traverse board was used to plot the course taken by the ship which had to progress by a series of tacks into the wind (**24**). Pegs are attached to the board by string, and these are pushed into different holes that indicate the compass bearing and the time steered on that bearing. The time was measured by a sand-glass in half-hours. At the end of the watch, the mean course and distance sailed could be worked out. Speed through the water also needed to be known, and this was taken by the log-line and charted by pegs in the rectangular part of the board.

The helmsman's traverse board is mentioned in 1528 as part of the navigator's set of necessary instruments. It continued to be used by coasting vessels in northern waters well into the 19th century. Made of wood, and measuring about 12 x 8 inches (30 x 20 cm), these objects were part of the ship's equipment that would not be thought important to keep when the ship was broken up.

Few have survived, and those that have are generally of the 19th century, although dating is difficult. Modern reproductions are quite common.

Marine Chronometer

The problem of finding the longitude at sea was solved in principle, though not in practice, by the 16th century. Two rival methods were discussed, the keeping of the time at the port of departure by a clock, and the measurement of the angular distance of the Moon against certain prominent stars compared against a set of standard tables. This method became known as the 'lunar distance' method, and was to be accomplished by using a cross-staff. The technology was not sufficiently advanced until the mid-18th century for this method to be of use, and even then it depended on long calculations, and on clear sightings of the Moon and stars.

The other method of finding longitude by using a clock, and comparing its time with a measurement of the local time at the ship's position, also depended on technological advance.

The first possible clockwork mechanism – a triumph of applied physics – was John Harrison's marine chronometer of 1735. He made two more of these large clocks, designed to be independent of the ship's motion and of temperature changes, in 1737 and 1757 (**23**), and, finally, a fourth in the form of a large pocket watch, 5¼-inch (13.5 cm) diameter, in 1759. It was this chronometer that set the pattern for all subsequent ships' clocks. These beautiful mechanisms may be seen working in the Old Royal Observatory, Greenwich.

John Arnold and Son, and Thomas Mudge Senior and Junior, were significant names in the early period of chronometer making. By the 1790s, the East India Company was using these new navigational aids, and Royal Navy officers would buy their own. The breakthrough in quantity production came in the late 1790s. By 1820, two firms, John Arnold and Thomas Earnshaw, had produced over 2,000 chronometers, and another English firm, Barraud, produced about 1,000. The

only competition came from France, where Le Roy was a pioneer, and was followed by Berthoud.

The industry, which was a craft of individuals, not a factory-based one, was active until about 1840 to satisfy the market for chronometers, but then the demand slackened because the instruments were so well made that they did not wear out – some are even perfectly satisfactory today. In 1840, the firms of Arnold and Dent were producing 60 box chronometers a year which sold for around £40 each.

From the 1850s and well into the 20th century, the two dominant English firms were Thomas Mercer and Victor Kullberg. It is not generally realised that these firms produced virtually the whole supply of marine chronometers for the English market and for a considerable part of the market abroad. Kullberg, for example, would sell a chronometer to Frodsham & Keen of Liverpool for £25; it was then retailed for £38, with, of course, the retailer's name on the dial. Similarly, the name on a Mercer instrument was often that of the retailer, and this applies equally to foreign retailers. Again, a chronometer of about 1840, signed by John Arnold of London, has the retailer's label in the lid – Peter Walther of Baltimore, United States. Although the maker's name and serial number may be found on the movement, these were sometimes changed, so there is no absolute guide as to the original name unless a firm's register book survives.

Other, smaller firms in England by the end of the 19th century were Johannson, Usher & Cole, and Dent. In France, real makers were Le Roy and Nardin, although their output was small. It is estimated that in 1889 the total output of British firms was 300 a year, whereas in France it was 40.

Early chronometers were placed in eight-sided mahogany cases, with a glass top. From about 1800, however, the form was a rectangular box, with brass binding, and a glass top. The instrument is mounted on gimbals. The mechanism is a masterpiece of craftsmanship, but it needs considerable study to understand the intricacies of the workings.

The National Maritime Museum, Greenwich, England, keeps a register of chronometers. This is an important register, because the design is constant, with the result that some chronometer designs may be older than one would think, and others more recent, perhaps even from the 1930s.

In 1997, late 19th- and early 20th-century examples were being sold for between £500 and £2,000, although earlier ones with a famous name would cost more.

Dip Circle

A magnetic device invented by the Elizabethan navigator and instrument maker, Robert Norman, is the dip circle, or dipping needle. This is a magnetic needle that measures the vertical component of the Earth's magnetic field, whereas in the compass it moves in a horizontal plane. Norman discovered the effect of dip in 1576. Then, it was thought that the angle of dip could find the latitude more easily than other means, but this proved false. The dip circles found today are not navigators' equipment, but those of scientific explorers, who study the way the Earth's magnetic field varies from place to place. Captain Cook took with him on his second voyage a dip circle made in 1772 by Edward Nairne, then one of the foremost of London's scientific instrument makers. Dollond, Troughton and other makers were making them at the end of the 18th century.

Known as Hedley's miner's dial; signed: *Davis & Son, Derby*, No. 216; c. 1880. John Hedley, an Inspector of Mines, commissioned in 1850 the firm of Davis & Son to make a dial to his own design for use in underground tunnels. (PC)

SURVEYING INSTRUMENTS

Wandering tribal peoples do not need to survey land; it is not until settled, agricultural communities develop that the division of land into definable plots becomes important to people. At first, the boundaries have to be fixed to mark agreed division between neighbours; then the areas, location and value are deemed necessary for taxation and for property rights.

Archaeologists have shown that some form of surveying existed in the river valleys of the Tigris and Euphrates, and of the Nile, before 1000 BC. The instruments appear to have been inexact, composed as they were of cords, rods and simple sighting devices.

The earliest book known to have been written on surveying is by Hero of Alexandria, a Greek engineer and scientist who was extremely successful in about AD 100. His book, *Treatise on the Dioptra*, gives the basic principles of the art of surveying, and describes a levelling instrument, the dioptra. The Romans used a similar instrument, a T-shape with open sights and a plumb bob, and also a level which consisted of a long, shallow trough filled with water. Still water always presents an exactly horizontal surface.

Roman techniques continued to be used until about 1500, when increased wealth and an enlarged population encouraged surveying as a profession. In England, following the dissolution of the monasteries by Henry VIII in 1539, the new owners favoured by the king needed to have their land demarcated.

Doubtless it was this activity that encouraged one Leonard Digges, a graduate of Oxford University, to invent a form of theodolite consisting of a horizontal circle divided into 360°, with a semicircle at right angles above it that could measure angles of elevation. His design was published in 1556, and there are two theodolites in existence, made in London by Humphrey Cole, bearing the dates 1574 and 1586, that put into practice the design of Digges. A superb theodolite was constructed in about 1600 by Erasmus Habermel, the Prague instrument maker retained by the Emperor Rudolf II.

Apparently, however, Digges was not the first to think of putting a vertical circle above a horizontal one, in effect two astrolabes, suitably mounted and modified. The German Waldseemüller had his invention illustrated in a book published in 1512, much earlier than the theodolites of Digges's design.

A new instrument of considerable practical and scientific importance was the magnetic compass, known to have been in use in China during the 11th century for cartographic surveys. To Alexander Neckham we owe the earliest reference to the compass for navigation in his *De Naturis Rerum* (1180). The magnetic compass became established at sea long before it was accepted on land.

By the early 16th century, however, an astrolabe had been manufactured that incorporated a compass; this was added to the circumferentor, an angle-measuring instrument that was a development of the astrolabe, used horizontally and not in its normal position. The compass was incorporated as a matter of course in the design of many later 16th-century theodolites.

The compass needle points to magnetic North, not to the geographic North Pole. The difference is known as the declination (or deviation) of the magnetic needle. Its value depends on the location and on the year. At London, the needle can move through 35° in 240 years (see Table II). A knowledge of magnetic declination can help in dating, when it is marked on a compass or on a sundial. In 1997, for the latitude of 50° North at the meridian of Greenwich, the value of the declination was 3° 6 minutes. The annual decrease, as calculated at present, is 9 minutes.

Levels

Levelling is the art of finding a line parallel to the horizon at one or more stations, to discover

TABLE II

The magnetic declination near London

Date	Declination	Date	Declination	Date	Declination
1580	11° 15′ E	1817	24° 36′ W	1871	20°10′ W
1622	6° 00′ E	1818	24° 38′ W	1872	20°00′ W
1634	4° 06′ E	1819	24° 36′ W	1873	19°58′ W
1657	0° 00′	1820	24° 34′ W	1874	19°52′ W
1665	1° 22′ W	1858	21° 54′ W	1875	19°41′ W
1672	2° 30′ W	1859	21° 47′ W	1876	19°32′ W
1692	6° 00′ W	1860	21° 40′ W	1877	19°22′ W
1723	14° 17′ W	1861	21° 32′ W	1878	19°14′ W
1748	17° 40′ W	1862	21° 23′ W	1879	19°06′ W
1773	21° 09′ W	1863	21° 13′ W	1880	18°57′ W
1787	23° 19′ W	1864	21° 03′ W	1881	18°50′ W
1795	23° 57′ W	1865	20° 59′ W	1882	18°45′ W
1802	24° 06′ W	1866	20° 51′ W	1883	18°40′ W
1805	24° 08′ W	1867	20° 40′ W	1884	18°32′ W
1806	24° 15′ W	1868	20° 33′ W	1885	18°25′ W
1809	24° 22′ W	1869	20° 26′ W	1900	16°52′ W
1812	24° 28′ W	1870	20° 19′ W	1910	16°03′ W

how much one plane is higher than another – an important skill required when cutting canals as this task demands accuracy, and allowance has to be made for the curvature of the Earth. For building work, simple levels are adequate.

Erasmus Habermel of Prague, in about 1600, included with a set of beautifully made surveying instruments, a water level. For convenience, however, a spirit level consisting of a long glass tube containing coloured spirits of wine was preferable. An early 18th-century example exists that is set in mahogany and is about 2 feet (60 cm) in length. Simple spirit levels with a pair of sights, fitting on a staff, were called drainage levels in the 19th century. Even with telescopic sights, a bubble tube is essential, and theodolites carry a pair of levels set at right angles to each other.

The oldest form of the modern surveyor's level was devised by Jonathan Sisson, a renowned mathematical instrument maker working in London during the second quarter of the 18th century. The pattern is known as the Y-level, because the telescope is supported in Y-shaped bearings, with brass straps over the top to hold the tube tight, enabling the telescope to be reversed to check the setting. Below the telescope is the bubble tube, and in the base, above the fixing at the head of the tripod, is a large magnetic compass.

Improvements were made to the level by Jesse Ramsden in the last quarter of the 18th century. A hundred years later, the Y-level was no longer popular in Britain, but continued to be used in the rest of Europe and in the United States. Although a good instrument, it had too many adjustments for use in difficult terrain such as is found in Africa and India.

The Y-level was superseded by the dumpy-level, in which the telescope could be fixed, because lenses were accurately centred by the 1840s. The civil engineer William Gravatt produced his design for the dumpy to be used during the railway boom of 1848. Compact and robust, it had a large objective lens in the telescope, and a cross bubble in addition to the bubble tube on the telescope. The Y had a telescope about 20 inches (50 cm) or more in length; the dumpy had one 12 inches (30 cm) long.

Levelling Staff

A level is used in conjunction with a levelling staff to measure heights. It consists of a rectangular wooden tube with two further sections that telescope into the outer one. Closed, the length is about 5½ feet (1.65 m); open, 14 feet (4.2 m). One face is marked out in feet (red) and tenths (black). A popular type was that invented by Thomas Sopwith in 1838, but surveyors tended to commission their own designs, so there are variations among those found today .

Barometer

For measuring great altitudes in mountains, where the ordinary operations of levelling are impossible, the barometer was used. The air pressure varies with height and with temperature, so it is possible to work out altitudes on mountains, or in air-balloons, with a barometer and thermometer. In about 1800, 'mountain barometers' were made and sold by W. & S. Jones of Holborn, London. Such barometric measurements were greatly improved after the invention, in 1845, of the aneroid ('without fluid') barometer by Lucien Vidie of Paris. Pocket aneroids, the size of a watch, were made from 1860 by Negretti & Zambra of London, and were able to measure heights of up to 20,000 feet (*c.* 6,000 m). A temperature-compensated surveyor's aneroid, 4½ inches (11.5 cm) in diameter, was sold around 1900 by W. F. Stanley & Co., for altitudes and depths of mines, that was claimed to give results accurate to one yard (90 cm).

Theodolite

Invented in the mid-16th century, the theodolite was not developed significantly until the 18th (**27**). Jonathan Sisson's first theodolites were traditional, with plain sights, but he made the important step of substituting a telescopic sight. Jesse Ramsden improved degree-scale division and by 1800 the instrument was almost in its final form. It is regarded as the most important of the surveying instruments, because it can measure, at the same time, both the horizontal angles between two points and their angle of elevation. It consists of a horizontal divided circle, the diameter of which gives the size of the instrument, e.g. 6-inch (15-cm), probably with small microscopes to read the verniers; a compass; a fixed telescope (not always fitted) below the compass; and a semicircular divided arc with the curve uppermost on earlier instruments, downwards on later models, surmounted by a telescope with bubble tube, as on a surveyor's level. A transit theodolite has a complete 360° vertical circle. With the surveys of the United States, Africa and India, and railway construction in all parts of the

world, during the 19th century, there were many variations in constructional detail of the instrument, and special adaptations for such operations as laying out railway curves.

From the time when the Enclosure Acts brought large areas of land under cultivation by private land owners, throughout the Industrial Revolution, with its road, canal and railway building, and the extension of similar projects into vast areas of the Empire, the demand on the British market for surveying instruments of all kinds was enormous (**30, 31**). Therefore it is not surprising that levels, theodolites and other surveying precision instruments were made by the first rank of English makers. Sisson and Ramsden have already been mentioned, and one should add George Adams, Thomas Jones, William Cary, Dollond, Edward Troughton (after 1826, Troughton & Simms), Negretti & Zambra, Elliott Brothers, J. H. Steward, W. F. Stanley and T. Cooke & Sons, York (after 1922, Cooke, Troughton & Simms Ltd.). European makers of standing were F. W. Breithaupt & Sohn, Kassel (awarded a prize medal at the 1851 Great Exhibition); Claude Langlois and Etienne Lenoir of Paris; Beaulieu of Belgium (another 1851 prize medallist); and W. Schenck & Co. of Berne. In the United States, the Rittenhouse brothers of Philadelphia made surveying instruments in the 1780s.

Graphometer

This came to be regarded in the late 19th century as satisfying the need for a simple, strong, portable, but inexpensive and accurate instrument for preliminary surveys (**26**). It could work in the horizontal and vertical planes.

Invented by Philippe Danfrie of Paris in about 1597, the instrument consisted of two alidades, or sighting rules, one fixed to a semi-circle divided in degrees, and the other movable over the scale of degrees. The whole was mounted on a tripod by means of a ball and socket joint. A magnetic compass was usually included. The graphometer was more popular in Europe, particularly France, than in Britain, although there are English-made instruments. Examples are signed by P. Danfrie, Vernier (1630), Pierre Sevin (1665), Choizy (1667), and the 18th-century makers N. Bion, Chapotot, Maquart and Langlois.

Circumferentor

A graphometer can be made with a full circle rather than a semicircle, and may be confused with the circumferentor. The original form of Digges's theodolite, it used open sights that were eventually replaced by telescopic sights. The compass box over the axis was made longer, and the superstructure was removed so that the needle could be read accurately by placing the eye directly above it; but the two pairs of horizontal sights were retained. These modifications produced the circumferentor from the old theodolite, existing alongside the telescopic theodolite (**25**). By the end of the 18th century, the circumferentor consisted of a pair of sights on the North–South axis of a large compass, of which the South end of the needle took the bearings. The compass rose is divided contrary to the usual way, East being where you would expect to find West. Thus, if you sight to the SE, the needle has to settle to give a reading of 45° SE, which would read SW on a normal compass rose.

By the early 19th century, theodolites were so improved that the circumferentor, along with the graphometer, went out of favour in Europe; but it remained important for taking surveys through woods and uncleared ground, as in America. It was also used in coal mines, and the term 'circumferentor' became an alternative to miner's compass. Here a large compass is specially important, because it may be the only way angles can be taken in low and tortuous mine tunnels. The compass is on a bar with folding sights at each end, and a pair of spirit levels.

Circumferentors were made in Italy, France, Holland and England. Those dating from the 18th century may bear the names of John Worgan, Thomas Heath, John Bennett and George Adams, all of London; and Butterfield in Paris. Mining compasses were made by W. & S. Jones, Dollond, William Cary, Negretti & Zambra and William Stanley. Provincial retailers' names are also found on standard products from London makers, or on instruments imported from Paris.

Plane-table

Surveying by the plane-table (before 1830, plain-table) is rapid, as every angle taken in the field is plotted directly, every distance is plotted to scale at the time, and all distances are laid off by scale and compasses. The table is a smooth board, of beech or mahogany, measuring about 20 x 15 inches (50 x 37 cm), with an outer edging that can fit over the board rather like the frame of a picture. This frame holds down a large sheet of paper placed over the board, and keeps it taut. Underneath is a brass socket which fits on to the top of a wooden tripod. On one side projects a magnetic compass box. On the top of the board is placed the alidade (index, or sighting rule), a sort of brass ruler, about 20 inches (50 cm) long, with one edge chamfered. At each end is a brass slit sight, and engraved on the upper side of the rule are a few scales (perhaps 1:10; 1:50; 1:100) for plotting ground measurements with the aid of a pair of compasses.

In use, the table is positioned so that the magnetic needle is due North, and then the index is used to sight various features, and their distances are measured; consequently, angles are drawn directly from the edge of the index, and distances are drawn to scale. By removing the table to other sites, a complete survey of a small region can be made in one step. Rainy weather would be a serious disadvantage, but was often predictable and therefore avoidable, as in India, where vast areas were covered by means of plane-tables in the late 19th century.

The plane-table was first referred to in English by Cyprian Lucar in 1590, in a text that described surveying by the use of the table alone; and it continued in use up to the 20th century. Nearly all mathematical instrument makers offered plane-tables, but they could also be made by any carpenter, and the compass could be bought from a ship's chandler, or elsewhere.

They are rare as collectors' items since, being made of wood, they may have decayed, or been thought of little value. Examples of plane-tables from the 18th century that exist in collections bear the names of George Adams, Senior or Junior, of London, and Langlois of Paris.

Gunter's Chain

This is used for measuring length on the ground. It is made of iron, so that it will keep its length, which a cloth tape will not. There are 100 links in a chain, the length of which is 22 yards (c. 20 m). The chain was developed in 1620 by Edmund Gunter, professor of astronomy at Gresham College, London. He was in favour of decimalisation, and a square chain is one-tenth of an acre (1 acre = 4,840 sq. yards). The chain, however, is based on a much older measure, known variously in different parts of Britain as the rod, pole or perch, which was 16½ feet (c. 5 m) in length, or one-quarter of a chain. Ten chains equal 1 furlong, which is one-eighth of a mile. Each 10th link is marked by a piece of brass notched for identification. By the mid-19th century, there were other chains in use, including lengths of 100-feet (30 m), used in North America, 50-feet (15 m) and 20-metres, with centimetre links. The product of wrought-iron smiths, the surveyor's chain is difficult to date unless there happens to be a maker's mark recorded somewhere.

Cross-staff

The cross-staff used by the surveyor must not be confused with the seaman's cross-staff, or fore-staff. The surveyor uses a brass or box-wood cross-sight on the top of a pole, the sightlines being exactly 90° to each other. It is used with a chain to measure 'off-sets'. When a chain is run out in a straight line, features on either side, such as a tree, river bend or building, can be drawn on the plan quite readily by measuring the distance from the chain line to the feature when the distance is taken exactly 90° from the chain. The cross-staff enables this to be done with some accuracy, and a simple and cheap survey can thus be made.

Optical Square

The principle is an old one, but the compact cross-staff dates from the 18th century, and was used by surveyors all over Europe. An improvement introduced in the 19th century is the optical square, where a half-silvered mirror at 45°, set in a small drum-shaped brass box, enables the surveyor to see the end of the chain line and the feature at right angles superimposed in the mirror, giving a more accurate starting point for the off-set. Sometimes the instrument is to be found with two pairs of cross-sights, so making the case octagonal, and a magnetic compass may be fitted at the top.

Examples found today are likely to be from the later 19th and early 20th centuries, and may bear the name of W. F. Stanley.

Surveyor's Wheel

The surveyor's wheel goes by several different names; in the 18th century it was generally known as a 'waywiser', and in the late 19th century as a 'perambulator'; it is also referred to as an 'hodometer', from the Greek, 'way measure'. A type of instrument dating from Roman times, it was reintroduced in the 17th century, and is used for measuring roads. On pavements and asphalt roads it is reasonably accurate, and is more useful than a chain in conditions of traffic. The outer rim, or tyre,

may measure 36, 72 or 100 inches (90, 180 or 250 cm), giving wheel diameters of about 11½, 23 and 32 inches (29.5, 59 and 82 cm) respectively. The revolutions of the wheel are recorded on a large dial mounted in a box, either adjacent to the axle or below the handle. The dial with two hands records yards, poles, furlongs and miles. The frame is nearly always of mahogany, and the tyre of iron or hard brass.

Made by specialists, probably makers of coach wheels, the signatures of the usual instrument makers are found engraved on the dial plates, for it was they who sold waywisers. London tradesmen include Heath, Heath & Wing, Martin, Adams, Dollond, Cary and W. & S. Jones. Late 19th- and 20th-century versions employ bicycle-type wheels, W. F. Stanley being a typical maker.

Hodometer

An associated instrument is the hodometer (also written odometer) which was fixed to a carriage wheel to record distance travelled. German and Dutch versions usually have three separate dials to record the measures.

Pedometer

This is a similar instrument to the hodometer, used to find out roughly the distance covered while walking (**28**). It was patented in 1831 by William Payne, and has the appearance of a watch with a single hand. The dial is normally marked 1 to 12 like a watch, so it may be confused with a timepiece. Inside, a weighted arm on a ratchet clicks up each step.

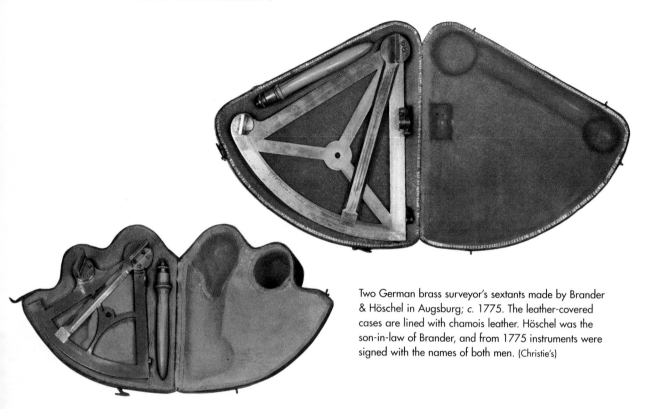

Two German brass surveyor's sextants made by Brander
& Höschel in Augsburg; *c.* 1775. The leather-covered
cases are lined with chamois leather. Höschel was the
son-in-law of Brander, and from 1775 instruments were
signed with the names of both men. (Christie's)

Pocket Sextant

The pocket sextant is a compact instrument devised by Edward Troughton in about 1800. It is also called a box sextant, and in its early years was known as a snuff box sextant, because of its size: only 2½ or 3 inches (6.5 or 8 cm) in diameter, and 2 inches (5 cm) deep. It was a great improvement on the cross-staff for setting 90° exactly, as, with a half-silvered mirror, the images of the chosen two points are made to coincide. Surveyors came to regard the pocket sextant as essential for use in trigonometrical observations, because any angle in the vertical or horizontal can be taken. A professional man wrote in 1888, 'This is an instrument, without which no surveyor should go into the field.' The small size,

compared with the seaman's sextant, was facilitated by the refined circle-dividing techniques brought into use by Jesse Ramsden, and improved by Edward Troughton. His tiny scales had to be read through a small magnifying glass.

Prismatic Compass

Another instrument regarded in the 19th century as essential equipment was the prismatic compass; since it could take bearings, and be used in traversing, it was the only reliable means of determining magnetic North in the absence of a theodolite. Compasses have been included as part of surveying equipment since the 16th century, and pocket compasses, varying in size from 2 – 6 inches (5 – 15 cm), have been separate instruments from that time to this. The prismatic was the first accurate compass – readable to a third of a degree. The name refers to a right-angled prism attached to the backsight that can be positioned over the rim of the compass card so that the degrees can be read while sighting the point of observation. The card is divided into 360°, which figure has to be inscribed on the South point, because that is where the eye and prism are positioned when looking due North. Also, the figures have to be printed in reverse, because of the reflection in the prism.

The prismatic compass was invented, and patented, in 1812 by Charles Augustus Schmalcalder, a mathematical instrument maker in the Strand, London. At the end of the patent, other makers copied the design, for example, Troughton & Simms, and Cary, and it became an essential instrument for walkers, explorers, and army officers.

Three small surveyor's sextants with silver degree scales; c. 1800. *From the left*: Signed: *Troughton, London*; *Ramsden, London*; *Berge, London*. Edward Troughton and Jesse Ramsden were two of the most expert at dividing arcs. Matthew Berge was employed by Ramsden. (Sotheby's)

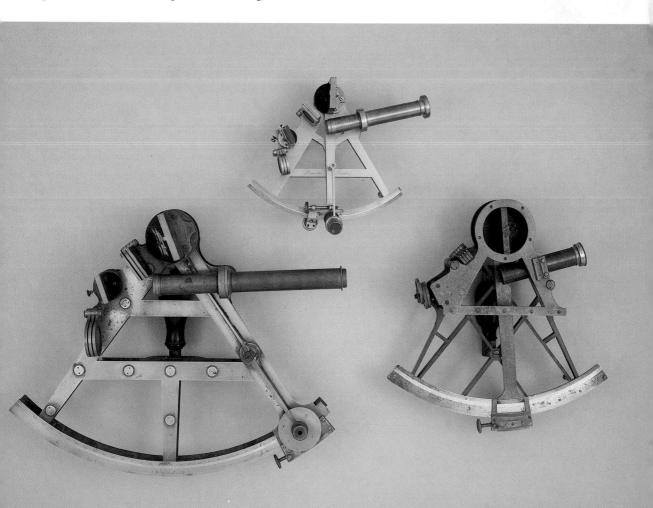

German trigonometer made in satin wood and brass, by Georg Brander of Augsburg; c. 1765. With adjustable, calibrated arc and plumb line, the instrument could be used as a clinometer or with a plane-table. (Christie's)

Clinometer

Sometimes the prismatic compass would incorporate in its brass case a clinometer, which measures vertical angles by means of a weighted wheel. This keeps its position while the case to which the sights are fixed is turned so that the point of observation is seen in the sights. The angle is read through the prism attached to the backsight. A clinometer scale is marked as well, giving the rise or fall in terms of inches per yard.

Another small, handy clinometer is the Abney level, invented before 1880 by William Abney of the School of Military Engineering,

Chatham. A small telescope is in a rectangular tube which has attached to it a semicircle divided into degrees and a clinometer scale. At the centre of the arc is a pivot with a bubble tube. The bubble can be seen through a hole in the telescope tube, via a mirror, so that the point sighted and the level bubble are viewed together. The angle through which the pivot has been turned is read off the scale.

Clinometer Rule

A simpler instrument is the clinometer rule, 6 inches (15 cm) long, of boxwood and brass, jointed like a carpenter's rule, with a small 90° quadrant at the join. Each arm has a bubble tube, and the top arm a pair of sights. An inclination table is inscribed on one arm, and sometimes a magnetic compass is incorporated. This is a useful instrument for geologists.

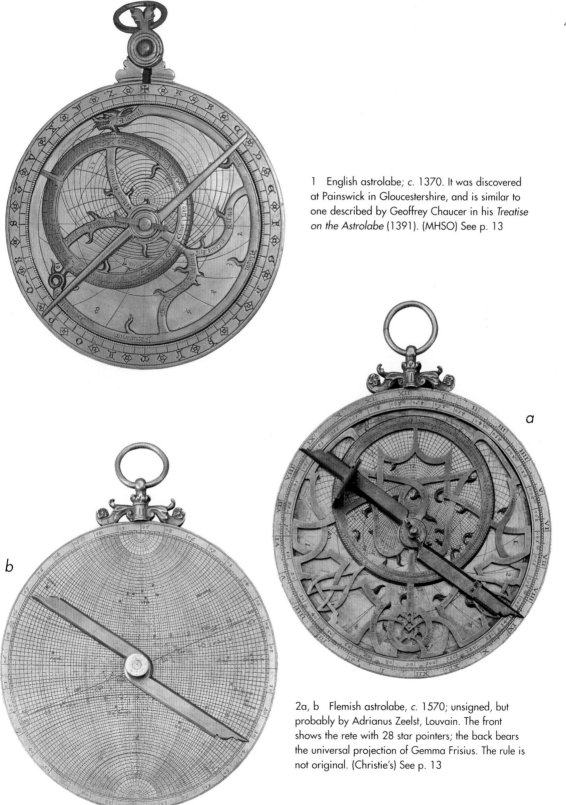

1 English astrolabe; *c.* 1370. It was discovered at Painswick in Gloucestershire, and is similar to one described by Geoffrey Chaucer in his *Treatise on the Astrolabe* (1391). (MHSO) See p. 13

a

b

2a, b Flemish astrolabe, *c.* 1570; unsigned, but probably by Adrianus Zeelst, Louvain. The front shows the rete with 28 star pointers; the back bears the universal projection of Gemma Frisius. The rule is not original. (Christie's) See p. 13

3 An Indo-Persian astrolabe in the Mughal tradition, unsigned, but probably produced in Lahore, *c.* 1700. There are five plates with projections for latitudes from 21° to 50°, and a rete for 32 stars. On the back is a shadow square, a quadrant of sines, and lines of solar declination. (Christie's) See p. 13

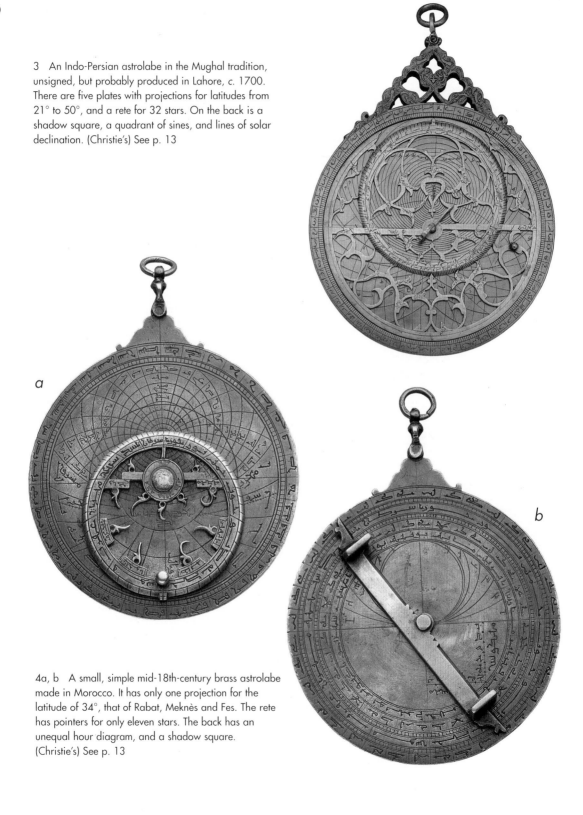

a

b

4a, b A small, simple mid-18th-century brass astrolabe made in Morocco. It has only one projection for the latitude of 34°, that of Rabat, Meknès and Fes. The rete has pointers for only eleven stars. The back has an unequal hour diagram, and a shadow square. (Christie's) See p. 13

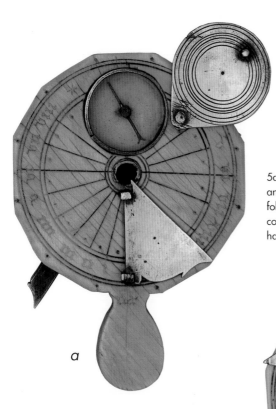

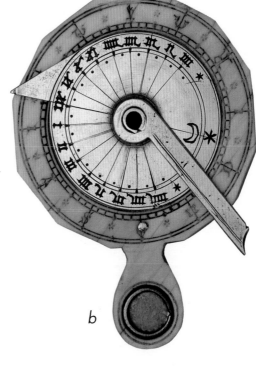

5a, b European horizontal dial and nocturnal in ivory and brass, probably early 16th century. The dial has a folding gnomon and a small magnetic compass with a cover. The nocturnal side has a calendrical scale; the handle has a small mirror. (Christie's) See pp. 17, 18

a

b

6 A rare example of an English diptych dial made in ivory, with a magnetic compass and a string gnomon for 52°. Inside the lid is a finely drawn, circular (2-inch diameter) map of England, unsigned, but by Charles Whitwell of London, *c.* 1600. (MHSO) See p. 20

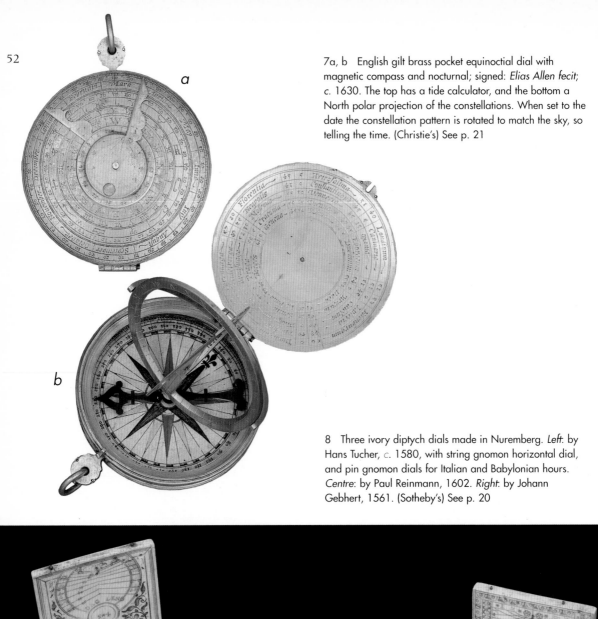

a

b

7a, b   English gilt brass pocket equinoctial dial with magnetic compass and nocturnal; signed: *Elias Allen fecit*; c. 1630. The top has a tide calculator, and the bottom a North polar projection of the constellations. When set to the date the constellation pattern is rotated to match the sky, so telling the time. (Christie's) See p. 21

8   Three ivory diptych dials made in Nuremberg. *Left*: by Hans Tucher, c. 1580, with string gnomon horizontal dial, and pin gnomon dials for Italian and Babylonian hours. *Centre*: by Paul Reinmann, 1602. *Right*: by Johann Gebhert, 1561. (Sotheby's) See p. 20

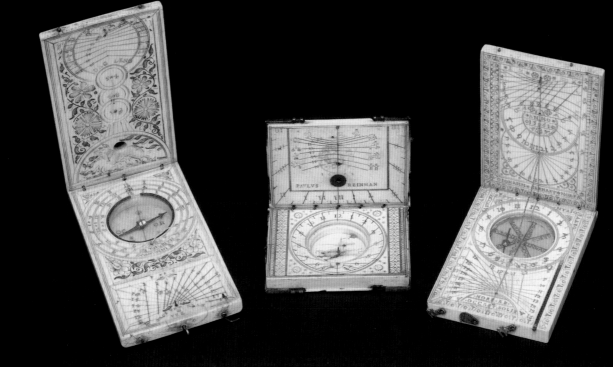

9 Dutch astronomical compendium in gilt brass, signed
and dated by a famous maker of Delft: ANTHONŸ
1666 SNEEWINS. The square plate is engraved with a
perpetual calendar on the underside, and above is a
magnetic compass and equinoctial dial. The cover is
a nocturnal with calendrical information. (Sotheby's)
See p. 21

10 Typical English mariner's nocturnal in boxwood,
dated 1679, bearing the unusual name of Gamaliell
Meggison. One side has the calendar and zodiac scales
with two pointers on a volvelle for the Little Bear (LB) and
Great Bear (GB). The long arm is to align on the
appropriate stars in either constellation to find the time
at night. The other side has calendar and 24-hour
scales, and a volvelle with the azimuths of several stars
and their declinations. (Christie's) See p. 17

11 English universal equinoctial ring dial of brass, signed: B. Scott Fecit. Benjamin Scott was working in London between 1712 and 1733, when he emigrated to St Petersburg to work for the Academy of Sciences. (Christie's) See p. 22

12 A standing universal equinoctial ring dial, signed: THO: HEATH LONDON; c. 1740. Such dials are simple and accurate, and when mounted with levelling screws, cross bubble levels, and a magnetic compass, can give the local solar time to one minute. Thomas Heath was working between 1720 and 1750. (PC) See p. 22

14   Danish portable brass equinoctial dial, signed: *Carl von Mandern fecit Hafnia* (Copenhagen); *c.* 1730. The hour plate can be adjusted for latitude, and the co-latitudes of eight Danish towns are engraved. The base incorporates a magnetic compass. Carl von Mandern was also a clockmaker. (Christie's) See p. 21

13   The gnomon of a monumental sundial at Jaipur, India, 1743. Known as the *Samrat Yantra* (the king of instruments), and built for Maharaja Sawai Jai Singh II, this is the largest sundial ever built. Its height is 90 feet (27.4 m) and its angle of 27° is the latitude of Jaipur. On each side are quadrants set in the equatorial plane that receive the shadow of the gnomon to tell the time. (GLET) See p. 11

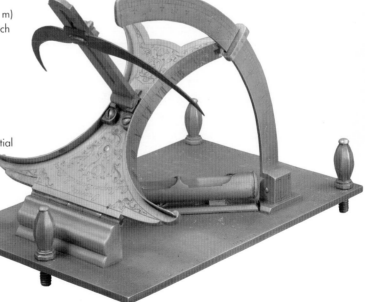

15   The crescent dial is a variation on the equinoctial dial, where the hour circle is divided in two and exchanged, forming a double crescent. The gnomon is also a crescent, the tips casting the shadow. Signed: Baradelle, Quai de l'Horloge du Palais, Paris; *c.* 1780. Also inscribed: *Inventé Par Jacques Baradelle AParis.* His working life was 1752 to 1794. (Christie's) See p. 21

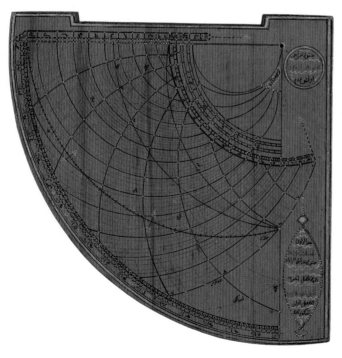

16 Universal mechanical equinoctial dial, signed: *Inventé et Exécuté par Hoyez à Amiens*; c. 1800. The large magnetic compass in the base, levelling screws and plumb line serve to orient the dial, and the latitude is set on a divided arc. When the dial plate is turned so that a spot of light from the Sun falls on the centre line of the long arm, the time is read from the hour disc by a geared pointer. (Sotheby's) See p. 22

17 The armillary sphere is used for teaching the elements of astronomy, and this English example, made in brass, dates from the mid-18th century. The small Earth globe at the centre was made by Nathaniel Hill in 1754. (Sotheby's) See p. 24

18 Made from lacquered wood, this astro-labe-quadrant is of Turkish origin, datable to c. 1860. It is signed by the maker: Dawudi. (Christie's) See p. 14

19 Two mariner's astrolabes recovered in 1973 off the coast of Florida from the wreck of the *Nuestra Señora de Atocha*, which sank in 1622. Both are of Portuguese manufacture, and are dated 1605 and 1616. (Sotheby's)
See p. 30

20 Portuguese mariner's azimuth compass; signed and dated: *Joseph da Costa Miranda afez em Lisbon Anno 1711*. The compass card is printed and coloured and is contained in a box with a glass lid. It is not the common ship's compass, but an instrument to measure the magnetic declination (known as variation to navigators). The bearing of the Sun at sunrise or sunset is taken through the side windows which have vertical wires to cast shadows on the compass. (WMHS)
See p. 34

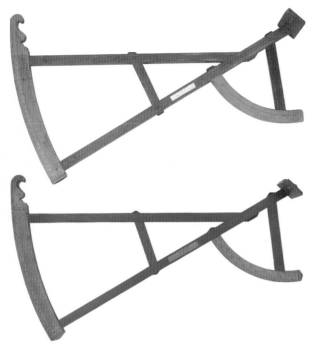

21 Two back-staves; mid-18th century. *Top*: signed: H. Gregory near the India House London [for] Jean Gremon. *Bottom*: signed: Thos. Greenough fecit 1753 For John Ahier. Thomas Greenough came from Boston, Massachusetts. (Sotheby's) See p. 31

22 This large octant of ebony and brass was made by Benjamin Martin of London; c. 1760. Mounted on a stand with a magnetic compass in the base. Usually thought of as a sea-navigator's instrument, octants were also used on land, as for example in plotting the line dividing New York State from Massachusetts. (PC) See p. 31

23 John Harrison's experimental chronometer No. 3, finished in 1757. It is a brilliant example of applied physics in its temperature compensation and self-oiling bearings, among other novelties in design. Harrison eventually won a prize for the invention of a reliable sea-going chronometer. (NMM) See p. 36

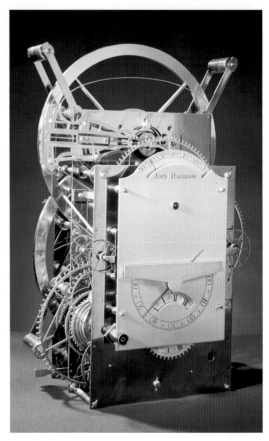

24 A very rare example of an English wooden traverse board, mid-19th century. Above is a rectangle with a series of peg holes: this is the log speed record. Below is a disc with 32 points of the compass, with a series of holes that record the direction sailed. Very few traverse boards survive since they were not the property of the navigating officer. (MHSO) See p. 35

25 This Dutch gilded brass and elaborately decorated Holland Circle is the finest known to exist. It is signed: Anthonius Hoevenaer Fecit Leydæ. This instrument is a form of circumferentor. There are four fixed cross sights, and a pair of sights on the alidade that rotates round the magnetic compass. A shadow square is incorporated, which is not necessary, but serves to show the derivation of the Holland Circle from the astrolabe. Hoevenaer was the first known instrument maker to a university department, matriculating at Leiden on 25 March 1683 as amanuensis to Professor Volder. (Christie's) See p. 43

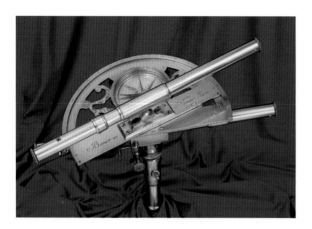

26 Popular among French land surveyors, this graphometer is signed and dated: *Bernier au Niveau, AParis, No.6, 1771.* (Christie's) See p. 42

27 This large geodetic theodolite is signed: *Nairne & Blunt, London.* It can be arranged to measure altitudes and azimuths or be set to the latitude of the station and measure equatorial co-ordinates. Edward Nairne and Thomas Blunt were in partnership between 1774 and 1793. (Christie's) See p. 42

28 This English pedometer is signed: RALPH GOUT LONDON No. 172. The mechanism is actuated by a chain, so that when the person wearing the instrument moves, every jerk makes the watch case fall against a spring and so pull the chain. Gout was granted a patent for the invention on 4 November 1799. (Christie's) See p. 45

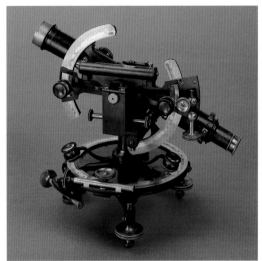

29 A fine French repeating circle by Gambey of Paris; c. 1830. The circle is counterbalanced and can be located in any plane; there are two telescopes. Readings on the silver scale are taken by four verniers that are viewed through microscopes. Henry Prudence Gambey was the most important Parisian instrument maker in the early 19th century. (PC) See p. 33

30 An English theodolite made to the design of George Everest, who as Surveyor General was responsible for the Survey of India, and after whom Mount Everest is named. It employs two vertical arcs of 80° instead of a full circle. Signed: Negretti & Zambra, London; c. 1880. (Christie's) See p. 42

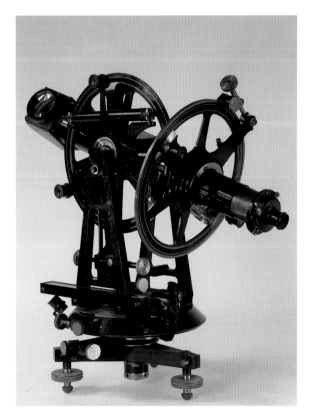

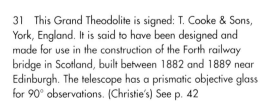

31 This Grand Theodolite is signed: T. Cooke & Sons, York, England. It is said to have been designed and made for use in the construction of the Forth railway bridge in Scotland, built between 1882 and 1889 near Edinburgh. The telescope has a prismatic objective glass for 90° observations. (Christie's) See p. 42

32 German-made proportional compass in gilded brass; *c.* 1600. Ratios are obtained by setting the points at one end to a line, fixing the central pivot (which moves in slots) to the desired ratio, when the other end of the instrument defines the length of the line at the new scale. Such instruments existed in the 1560s, and became more popular around 1600. They are still used today. (Christie's) See p. 82

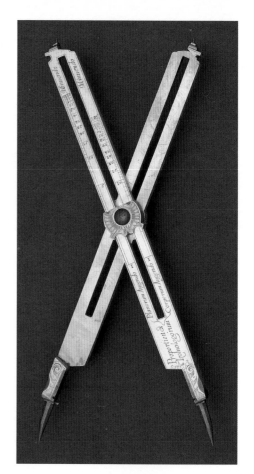

33 Drawing instruments made in Rome by Jacobus Lusverg (dated 1687, 1689) and Dominicus Lusverg (dated 1684, 1723). (PC) See p. 81

34 English gunner's callipers, signed: I. Rowley Fecit; *c.* 1720. It was made for artillery officers to gauge the calibre of guns and the diameter of shot to be used. John Rowley was working from 1698 till his death in 1728. He was Master of Mechanics to King George I. (Christie's) See p. 81

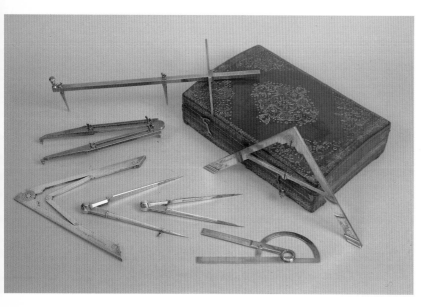

35 A comprehensive set of drawing instruments
produced in London, c. 1880. The circular protractor is
engraved: *L. CASELLA Maker to the Admiralty & Ordnance
LONDON*. Larger items include an ivory
sector, nickel-plated 20° set-square, eleven French curves
and a beam compass. (MHSO) See p. 81

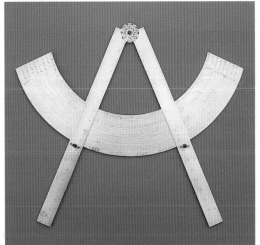

36 English brass architectonic sector; unsigned,
probably made by Thomas Heath, London; *c.* 1730. The
two arms form a sector, while the wide arc is
finely engraved with scales to calculate the elements that
embrace the five classical orders of architecture.
Designed by Thomas Carwitham, who published in
1723 a book on it with Thomas Heath. (MHSO)
See p. 84

37 French, late 19th century cased set of drawing
instruments in nickel-brass, steel and ivory. (Christie's)
See p. 81

38 An unusual set of Napier's bones made by
George Adams Senior of Fleet Street, London;
c. 1760. With this set, calculations are based on
the sexagesimal system used in astronomy.
(Christie's) See p. 87

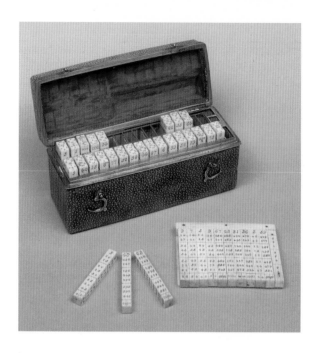

39 This set of geometrical models illustrates the
theorems of Euclid. Made by George Adams
Senior at the sign of Tycho Brahe's Head in Fleet
Street, London; *c.* 1760. (PC)
See p. 105

40 A large English compound microscope signed: John Marshall,
Ludgate Hill, London; *c.* 1695. John Marshall advertised this design in
1693, and continued to produce it until his death. The body is of lignum
vitae and pasteboard covered with coloured and gold-tooled vellum. The
sub-stage mirror is a later addition. (Sotheby's) See p. 93

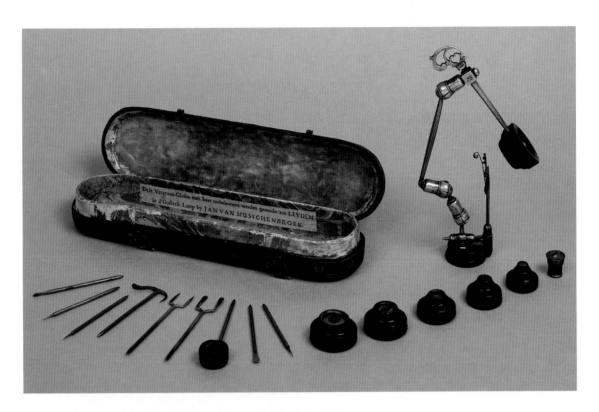

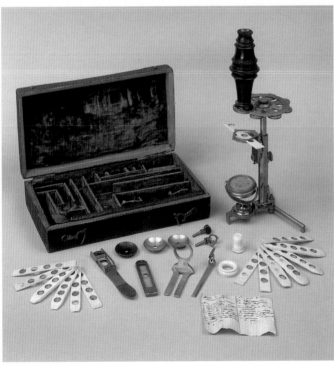

41 A Dutch set of simple microscope objective lenses in ebony mounts and brass specimen probes, together with the holder, pack into a box covered in gold-tooled leather; c. 1690. The box has the trade label of Jan van Musschenbroek, and the holder has his trademark: an Eastern Lamp with the crossed keys of the city of Leiden. The ball-and-socket joints are known as 'Musschenbroek nuts'. (Christie's) See p. 93

42 The 'New Universal Double Microscope' was made by George Adams Senior at the Sign of Tycho Brahe's Head, Fleet Street, London; c. 1750. Adams published this instrument in 1746. The body tube is of ivory stained black, and fits over a wheel containing eight different objective lenses. This is the first use of an objective changer or rotating nose-piece. (Christie's) See p. 94

43 This Culpeper-type microscope is signed: NATH. ADAMS OPTICIAN TO HIS ROYAL HIGHNESS FREDERICK PRINCE OF WALES FECIT; *c.* 1735. The body is of lignum vitae, with the outer tube covered in rayskin. It is unusual in having four legs, and the substage mirror is on a universal ball mount. Nathaniel Adams was apprenticed to Scarlett in 1722, and was free in 1730. Frederick was created Prince of Wales in 1729. (RMS) See p. 94

44 Signed: B. MARTIN LONDON on the draw-tube, this solar microscope has a mahogany plate to attach to a window shutter; *c.* 1760. The outer tube is covered in rayskin dyed green, and the brass tube holds the ivory sliders and the projection lenses. Benjamin Martin opened his retail shop in 1756. (Christie's) See p. 96

45 These two compound binocular microscopes are signed (*left*): Swift & Son, 81 Tottenham Court Rd., London W.C.; (*right*): J. Swift, 43 University St, London W.C. The firm moved from University Street to Tottenham Court Road in 1881. The binocular tubes are to the design of Francis Wenham, 1861. (Christie's) See p. 97

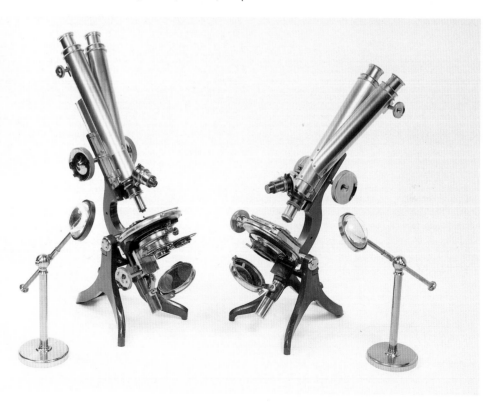

46 A fine example of the precision research compound binocular microscope of the 1890s, signed: W. WATSON & SONS, 313 High Holborn. LONDON. This model is described as the Van Heurck 'Grand Model' after Dr Henri Van Heurck of Brussels, who commissioned Watsons in 1891 to make a microscope to his specifications. It was popular for a quarter of a century. (Christie's) See p. 98

47 A group of eight telescopes. *Left to right from top*: opera glass *c.* 1840; opera glass *c.* 1750; binoculars *c.* 1890; opera glass with case *c.* 1780; reflector in case *c.* 1750; refractors *c.* 1790; *c.* 1780; *c.* 1830. (MHSO) See p. 98

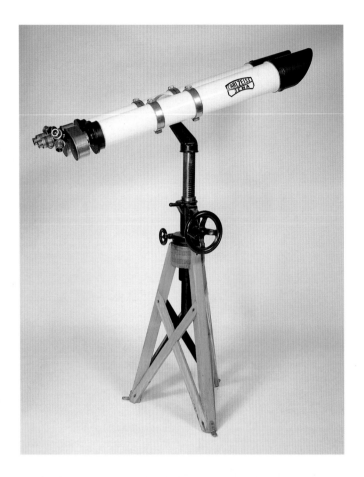

Opposite
50 German double lantern for dissolving views, signed: *A. Krüss. No 165. Hamburg*; purchased in 1861. Two identical lanterns fit one over the other, and each has special shutters so that one opens as the other closes. This gives an uninterrupted change from one slide to the other. (TM) See p. 100

48 German binocular telescope, signed: CARL ZEISS JENA; *c.* 1900. The eyepieces are on a revolver with a choice of powers, ×33, ×53, ×73. Such telescopes were used at observation points in mountains. (Christie's) See p. 99

49 A late 18th-century Sheffield plate and enamelled four-draw telescope, signed: DOL-LOND *LONDON*. The case is covered in red morocco leather. (Christie's) See p. 99

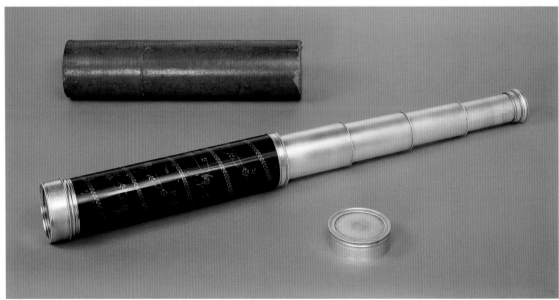

51 Two English lantern slides showing (*right*) the diurnal motion of the Earth, and the reason for night and day, and (*left*) proof that the Earth is round. Purchased in 1861. (TM) See p. 100

52 Viewer for stereoscopic pictures to the David Brewster lenticular pattern announced by him in 1849. Queen Victoria saw such a viewer at the Great Exhibition of 1851 and was amused by it, so creating a vogue, causing vast sales. This elaborately decorated viewer has a box to hold the stereo slides; *c.* 1865. (WMHS) See p. 101

53 A heliostat by R. Fuess of Steglitz, Berlin, a design introduced in 1879. A biaxial single-mirror instrument, well made but over-elaborate; *c.* 1890. Used in stellar spectroscopy and stellar photography to keep the image stationary during very long exposures. (PC) See p. 110

54 Three colour discs and machine to rotate them. Made in Italy, *c.* 1865. The variations in the length of the circumferential coloured arcs produce the appearance of different colour mixes when rotated. (PU) See p. 110

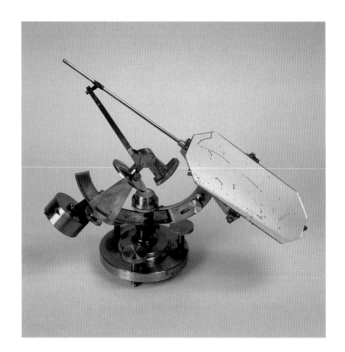

55 English double-barrelled air-pump of the pattern made *c.* 1700 by Francis Hauksbee. The carved stand is made in walnut wood. In 1709, Hauksbee published his influential *Physico-Mechanical Experiments*, which was translated into Italian, Dutch and French. (PC) See p. 106

56 A group of accessories for demonstrations with an air-pump. Made in Britain, *c.* 1900. (MHSO) See p. 106

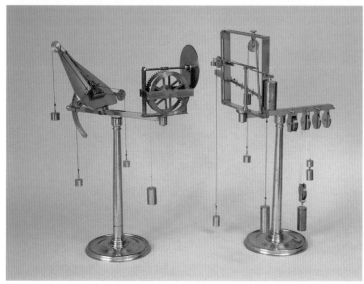

57 Set of mechanical demonstrations made by Benjamin Martin of London; *c.* 1765. Included are: various pulley blocks, balance, triple lever, rolling down an inclined plane, triangle of forces, worm drive. (PC) See pp. 103, 104

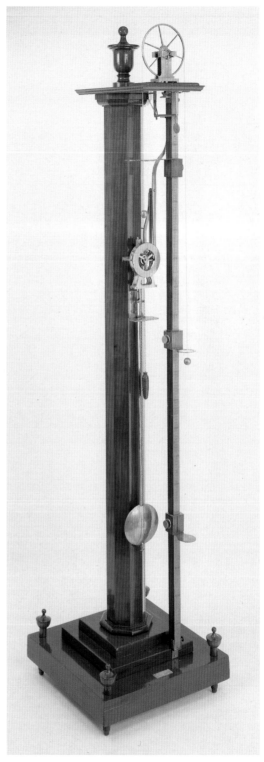

58 Apparatus to show that a body takes the same time to descend the diameter or chord of a circle whatever the length of that chord may be. Danish construction in mahogany; c. 1800. (SA) See p. 105

59 A fine example of a French Atwood fall machine, signed: BRETON FRÈRES Paris Rue Dauphine, 23; c. 1870. The Cambridge University mathematician, George Atwood, invented the machine and published on it in 1784. It is used to show the laws of motion uniformly accelerated or retarded, as well as bodies in uniform motion. (Christie's) See p. 104

60 Model of a James Watt beam engine, signed: WATKINS & HILL, 5 CHARING CROSS, LONDON; c. 1840. It is powered by burning charcoal, and was used in lectures at the Clarendon Laboratory, Oxford. (MHSO) See p. 105

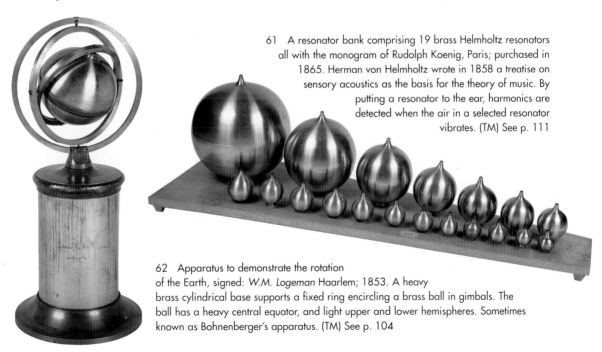

61 A resonator bank comprising 19 brass Helmholtz resonators all with the monogram of Rudolph Koenig, Paris; purchased in 1865. Herman von Helmholtz wrote in 1858 a treatise on sensory acoustics as the basis for the theory of music. By putting a resonator to the ear, harmonics are detected when the air in a selected resonator vibrates. (TM) See p. 111

62 Apparatus to demonstrate the rotation of the Earth, signed: *W.M. Logeman* Haarlem; 1853. A heavy brass cylindrical base supports a fixed ring encircling a brass ball in gimbals. The ball has a heavy central equator, and light upper and lower hemispheres. Sometimes known as Bohnenberger's apparatus. (TM) See p. 104

63 Six lamella horseshoe magnets, with keepers; before 1882. Made by Gebrs. Van Wetteren, Haarlem, The Netherlands. (TM) See p. 106

64 The cylindrical form of the electrostatic generator, signed and dated: W.C. COOPER; 1842. The accessories include an electric orrery, gold-leaf electroscope, Leyden jar, electric chimes and discharging tongs with glass handle. (Christie's) See p. 107

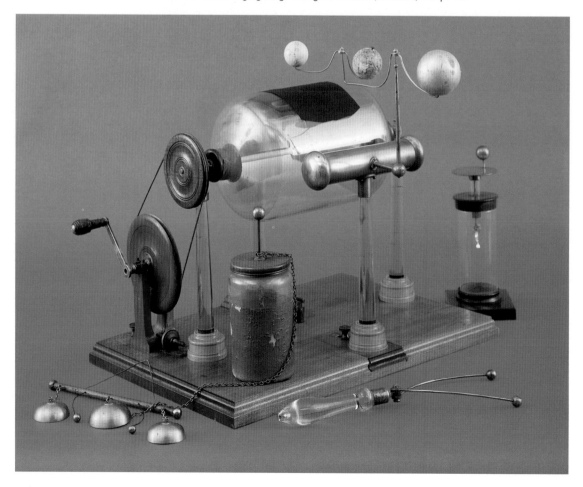

65   Battery of four Leyden jars; mid-19th century. (PU)
See p. 107

66   Wimshurst pattern
electrostatic induction generator,
with four Leyden jars and
adjustable spark gap at the top.
Signed: *Harvey & Peek, 56,
Charing Cross Road, London (Late
W. Ladd & Co.); c.* 1890.
James Wimshurst (1832–1903)
produced, in 1880, the most
powerful of the 19th-century
generators of static electricity.
(Christie's) See p. 107

67 A self-contained 'electric egg' with a bulb (the 'egg') in uranium glass and a Tate pattern vacuum pump below it. Signed: Newton & Co., 3 Fleet St., London; c. 1860. This apparatus demonstrates the aurora borealis. Thomas Turner Tate invented in 1856 a double piston air-pump that could achieve a higher vacuum than other pumps. (Christie's) See p. 107

68 Five discharge tubes of unusual design. Standing: gridiron Geissler tube; Crookes tube with fluorescent mineral; Hittorf double helix. Below: Crookes tube with seven bulbs; Geissler tube filled with fluorescent fluid. (Christie's) See p. 108

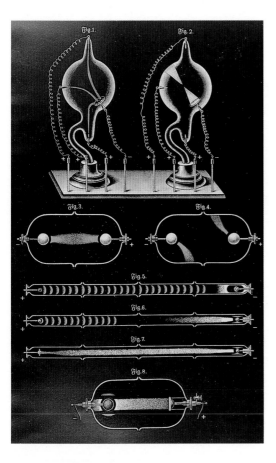

69 Electric discharges in rarefied gases. Plate VI in
J. Frick, *Physikalische Technik, oder Anleitung zu
Experimentalvorträgen*, 7th edn, part ii (Brunswick,
1907). See p. 107

70 A typically English 'birdcage' quadrant electrometer
invented in 1867 by William Thomson, later Lord Kelvin.
The fundamental instrument for measuring potential difference in volts. Capable of measuring 0.01 volts with a
range of 400 volts. (MHSO) See p. 109

71 Lord Kelvin's reflecting astatic galvanometer by
Elliott Brothers, 100 & 102 St Martin's Lane, London;
c. 1890. The curved bar over the top is a magnet to
counter local magnetic fields. William Thomson, a brilliant physicist, was responsible for laying the transat-
lantic telegraph cable in 1856; elevated in 1892 as
Baron Kelvin of Largs. (Christie's) See p. 109

72 *Left*: an angled barometer, signed: *Charles Howarth Fecit Halifax*; c. 1840. The angle was intended to increase the accuracy with which the pressure could be read on an extended scale. *Centre*: A mid-19th century sympiesometer, made to measure atmospheric pressure on board ship. *Right*: A bow-fronted thermometer, signed: *Dollond London*; c. 1790. The scales measure degrees Fahrenheit and Réaumur. (Christie's) See pp. 111, 113

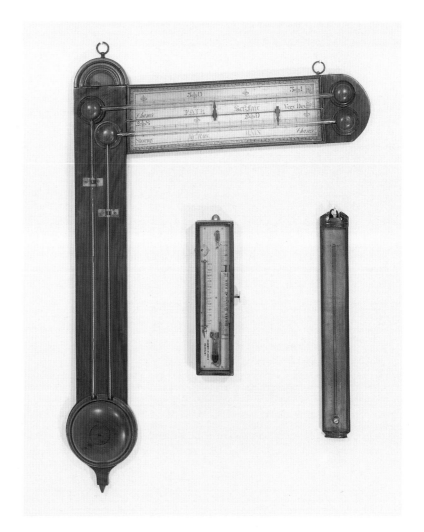

73 Unusual barometer with a fan-shaped indicator plate to one side. Inscribed with the name of a Yorkshire clockmaker: *THOMAS HARGRAVE*; c. 1785. (Christie's) See p. 113

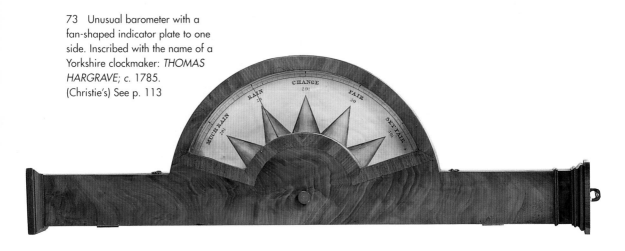

74 Portable medicine chest, Tirolese; late 17th century. The case is made in walnut, and the drugs are contained in screw-topped pewter pots or in glass bottles. (MHSO) See p. 130

75 Chemical balance, English, c. 1840, by George Knight & Sons, Cheapside, London. There are brass and platinum weights from 500 grains down to 0.1 grain. (MHSO) See p. 121

76   This small oil painting of about 1860 shows
electrical treatment of a woman sitting in an insulated
chair (the legs are on glass tumblers). The cylindrical
electrostatic friction generator is of the Edward Nairne
type. (MHSO) See p. 130

DRAWING AND CALCULATING INSTRUMENTS

ASTRONOMERS, navigators, surveyors, architects, draughtsmen and engineers have certain tools of their trade in common. All need to draw accurately, whether it is land boundaries on a map, co-ordinates on a star chart or the design for a steam engine, and all need to calculate, which in the present context means addition, subtraction, multiplication and the use of trigonometrical functions: basic stuff but time-consuming if done with pencil and paper. Hence, drawing instruments and calculating instruments were evolved.

Sets of Drawing Instruments

Drawing involves ruling a straight line, parallel lines, and lines at a given angle to another, as well as measuring parts of lines (**33**, **35**, **37**). For these purposes there are rulers, parallel rules that can extend to a required distance, and protractors. Measurements are normally to scale, for example, 1 inch may represent 1 mile on the ground (a scale of 1: 63,360), or a special scale is cut so that there are 4 chains to 1 inch (20 inches to 1 mile). A typical set of drawing instruments includes ruler and scale; parallel rulers; circular or semicircular pro-

tractor; plotting scales and rectangular protractor combined; a pair of compasses for drawing circles; spring bows for small circles; a pair of dividers for pricking off lengths; pens, pencils and similar attachments for the legs of the compasses. Sets may be extended by additional sizes of compasses and dividers, and certain other instruments, such as a sector or gunner's callipers (**34**).

Drawing instruments are produced in two qualities, for learners and for professionals. This distinction became more important during the 19th century when more technical colleges were founded and schools became more technically minded. The presentation of such sets varies; in the 18th century they were commonly in black, fishskin-covered cases, with flip-top lids, but could be in hinged boxes, perhaps rather ornate. In the 19th century the range of boxes was extended (**35**, **37**). By the end of the century, cases for professionals could be in mahogany or rosewood with brassbound corners, and one or more lift-out trays, lined with velvet, usually dyed blue. Flat leather-covered cases, with rounded corners and a bolt fastening, were made in France for carrying in the pocket, and imported and then copied by the British trade.

Compasses and dividers are generally of brass with steel points of triangular cross-section. The rulers and scales are of boxwood for cheapness, or brass, ivory and, occasionally, silver. Parallel rulers are of ebony, ivory or brass. In about 1880, Negretti & Zambra were supplying sets of mathematical drawing instruments made in German silver (a copper, zinc and nickel alloy) as being best suited for use in warm, damp climates where brass would quickly corrode. Large sets could include colours, palettes and brushes.

Additional instruments, sold separately, include a beam compass for large circles; three-legged or triangular compasses; an elliptical trammel, and ellipsograph for drawing ellipses; proportional compasses for taking ratios; rolling parallel rulers; circular protractors with hinged, folding arms with pricking points, tangent screw and vernier; set squares; sets of architectural and other curves, to provide a variety of ruling edges; jointed measuring rules; and a T-square and drawing board.

Draughting instruments are ancient in origin; compasses and dividers can be traced to Babylonian times; Leonardo da Vinci, at the end of the 15th century, sketched drawing pens and proportional compasses. In some museums there are superb examples of sets of drawing instruments made for courtly presentation by Italian and French craftsmen of the 16th and 17th centuries, with elaborately formed and decorated instruments in gilt copper or brass, placed in tooled leather cases.

Proportional Compass

The proportional compass is used to enlarge or reduce a drawing by having two arms and a pivot part-way along, so that the ends open out to different lengths, usually in the ratio 2:1 (**32**). It was known to the Victorians as a 'whole and half compass'. The version with a continuously variable ratio has two slotted arms with points at the ends of both arms. Connecting the two is a movable pivot, so that the ratio can be varied at will. The positions for certain ratios were marked: scale of lines for lengths; scale of circles to divide the circumference into any number of equal parts; scale of planes for areas; and scale of solids for volumes. The instrument is still made today.

The origins of the proportional compass remain obscure. It is depicted, however, in a portrait of Wenzel Jamnitzer made in *c.* 1565, while a claim has been made that it was designed by Federico Commandino of Urbino *c.* 1575. The earliest surviving example, *c.* 1582, is signed by Jost Bürgi, a Swiss instrument-maker.

There is sometimes confusion over the name of this instrument, called a proportional compass in English, a *compas de reduction* in French, *compasso di riduzione* in Italian, *Reduktionszirkel* in German, and *reductiepasser* in Dutch.

Pantograph

Change of scale is frequently required by draughtsmen, and to speed things up other new instruments were developed, in particular the pantograph, which was invented between 1603 and 1605 by the German astronomer, Christoph Scheiner, and greatly improved in 1743 by the Parisian craftsman, Claude Langlois. The railway building of the 19th century ensured a need for this device. It consists of four brass bars, jointed in pairs, one pair being twice the length of the other. Under the joints are small castors; one long bar has a tracing point, and a short arm bears a pen held by a sliding head that is set to the ratio required. On the other long bar is a pivot point in the form of a heavy brass disc.

Eidograph

The eidograph, an improved instrument for reducing or enlarging drawings, was invented in 1801 by a Scotsman by the name of William Wallace, who subsequently became professor of mathematics at the University of Edinburgh, Scotland. Although similar to the pantograph, any ratio could be taken between the limits of one to three, for example, 9:25.

Polar Planimeter

Areas on maps and plans have to be measured, and one laborious way is to place over the area in question a sheet of thin paper with a grid of small squares which then have to be counted.

The polar planimeter measures areas merely by tracing the outline. It consists of two arms, one with a pin to fix it to the board, and the other with a tracing point. At the joint is a small wheel that rotates as the tracing movement is performed, and the area is read off a dial.

This instrument was invented in 1854 by Jakob Amsler, professor of mathematics at the University of Schaffhausen in Switzerland.

Opisometer, Chartometer

The opisometer is a small device for measuring the lengths of roads, rivers, walls, etc., on maps. It is a milled wheel on a screw thread with a handle. The wheel traces the route, and is then wound backwards on the scale at the edge of the map. The chartometer is the same but has a dial and pointer to give the measure immediately.

Station Pointer

For surveying in new regions, and especially for hydrographic surveys, the station pointer is essential. It is a double-arm protractor, where two angles relative to a base may be laid off at the same time. For taking coastal soundings, the angles between three points on land are measured with a sextant, the two movable arms are set relative to the fixed arm and the instrument is placed over the chart. When the arms all match the features on the shore, the boat's position is fixed exactly, and the point is pricked on to the chart. The protractor's diameter varies from 5 – 12 inches (12–30 cm), and the arms are 12 – 15 inches (30–37 cm) long, with brass or wooden extension pieces. This instrument was the invention of the Admiralty Surveyor, Murdoch Mackenzie, who in 1774 published details in his book, *Treatise on Maritime Surveying.*

Sectors

Sectors were made with a variety of scales for use in calculation by navigators, surveyors, gunners and draughtsmen; at first sight they look like a jointed rule, and are made of brass,

Four English 6-inch sectors. *Top*: ivory, signed: THOMAS JONES 64 CHARING CROSS, *c.* 1850; *left*: ivory, signed: ELLIOTT BROS STRAND, LONDON, *c.* 1860; *right*: silver, signed: *Dollond London, c.* 1820; *bottom*: brass, signed: *J. Long London, c.* 1825. (PC)

wood, ivory or sometimes silver (**36**). The sector was in use in England by 1597 for gunnery, and Thomas Hood's design was made in brass with an arc of 150° and two arms which carried scales of proportion based on the principle of similar triangles. These scales were used in conjunction with a pair of dividers – a necessary accompanying instrument for all types of sector.

Galileo developed a sector between 1597 and 1599 for use as a general-purpose calculator that he called a compass; on the Continent it became known as the proportional compass. This has caused much confusion, because in Britain the proportional compass is a different instrument, described above. The European names for the sector are: *compas de proportion* in French, *compasso di proporzione* in Italian, *Proporzionalzirkel* in German and *proportionaalpasser* in Dutch.

Sectors were in general use for 300 years, but by 1866, W. F. Stanley, the instrument maker, could write, 'The sector is a kind of twofold rule, commonly supplied with a case of mathematical instruments, as a kind of established ornament.'

The sector is engraved with a number of scales of mathematical functions, and it can give similar information to that provided by a slide rule. French sectors differ slightly from English, having fewer scales, since they were intended for use in gunnery. The English instrument is essentially a draughtsman's aid, and, during the 18th or 19th centuries has the following scales:

L	line of equal parts, 0-10
C	line of chords, 0°–60°, used to protract an angle
S or Si	line of sines, 0°–90°
T	line of tangents, 0°-45°
t	line of tangents to a smaller radius, 45°–75°
S or Se	line of secants, 10°–75°
P or POL	line of polygons, 12–4, for inscribing a regular polygon inside a circle of a given radius

Some other scales may be engraved along the edges:

N or Num	line of numbers, 0–10 twice, used with a pair of dividers in the same manner as a slide rule for multiplication; it is Edmund Gunter's logarithmic scale described by him in 1624
R or Rh	line of rhumbs, 0–8, used for plotting a ship's course upon a chart; the scale is in points of the compass, where 32 is a complete circle
Lon	longitude, 60°–0°, used in navigation
La	latitude 0°–90° used in construction
Ho	hours, 0–VI of sundials

The French sector commonly has the following scales:

line of equal parts
line of chords
line of planes (areas)
line of solids (volumes)
calibre of pieces (size of gun barrels)
weight of shot
line of metals (six metals denoted by symbols – gold, lead, silver, copper, iron, tin), used for weight/volume measurements
polygons, 12–3

The use of these scales is unfamiliar in the 20th century: a good account can be found in Edmund Stone's translation of N. Bion, *Mathematical Instruments* (1758), reprinted in 1972 and possibly available in rare bookstores.

Abacus

The history of the abacus runs from the prehistoric era to modern times. The Greek, and later the Roman, abacus was no more than a convenient, flat surface on which pebbles could be placed. The word in Greek means a disc or table. The Latin phrase for reckoning accounts was 'ponere calculos' which means 'to place the pebbles'. From the word for pebbles derive our words 'calculate' and 'calculus', a branch of mathematics.

The Roman calculating board was divided into columns, headed with the letters which

Top: Chinese abacus, with a rosewood frame and beads on bamboo runners. Bought in San Francisco in October 1877. *Bottom*: Japanese abacus with a cherrywood frame, oak base, ivory divider, and bamboo beads and runners. Bought in 1929. Both abaci are traditional in form, and continue in use to the present date. The number set is 1,982. (MHSO)

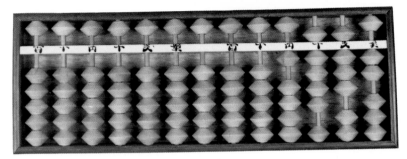

were used to designate numbers: M (1000), C (100), X (10) and I (1). Into each section could be placed anything between 1 and 9 pebbles. But it is difficult to add up 8 or 9 pebbles at a glance, so the intermediate numbers, 5, 50 and 500, were represented by 1 pebble placed on the line dividing two sections. These were designated V (5), L (50) and D (500). This meant that calculations could be worked with 1 pebble representing 5, and 4 singles.

The way in which the Roman calculating board was used is shown in contemporary illustrations, and the counting pebbles have been discovered on archaeological sites. There also exist a small number of portable bead calculators dating from the Roman period, which are now in museums in London, Paris and Rome. These devices are small enough to hold

in one hand, and consist of beads that are moved in vertical slots. The slots are divided into two sections horizontally; in the upper section is 1 bead for 5; in the lower, 4 beads to be used as singles.

So both the calculating board and the bead abacus existed in Roman times, and both continued in use for many centuries. In Europe, the board with counters was used by government departments and in commerce, and eventually had an interesting side-product in the game of shove-halfpenny, which must have started to pass an idle hour in the counting house, and is still popular in some English inns. Counting boards are rarely found, but the counters, known as jettons (French jetons) can occur. These were first imported to Britain from France in the 13th century, and from the 16th to the 18th were produced for use throughout Europe by craftsmen in Nuremberg. These tokens are made of brass, and usually have the head of the reigning monarch on one side and the name of the maker on the other.

The Roman bead calculator developed into the bead-frame abacus which is still used in China, Japan, Russia and Poland. A rectangular frame is set with parallel rods or wires, on which beads slide up and down. Chinese abacuses have rounded beads on bamboo rods, 2 in the upper section, 5 in the lower. Japanese examples, which first became common in the 17th century, have diamond-shaped beads, disposed with 1 in the upper section and 4 in the lower. Russian and Polish abacuses have wooden beads on wire rods, with 10 on each wire; the two middle ones, 5 and 6, are painted black. As well as examples of Far Eastern and Eastern European abacuses from the 19th and 20th centuries, it is possible to locate examples of bead frames made for teaching children in Britain during the Victorian and Edwardian periods.

It is interesting to note that modern mathematics and modern technology have brought back the principle of the abacus, after a period of pen reckoning.

Napier's Bones

The techniques of adding and subtracting were well established from antiquity, but by the 16th century it became necessary to develop the new technique of multiplication, for navigation and surveying. This was not possible with Roman numerals, but the introduction of Arabic mathematics made it feasible, although very laborious. John Napier, laird of Merchiston in Scotland, devised logarithmic tables and published his discovery in 1614. He wrote that the multiplication and division of great numbers is troublesome, involving tedious expenditure of time, and subject to 'slippery errors'. His tables reduced these difficulties to simple addition and subtraction, and won immediate recognition. In 1617, he published another book which included the technique of using the rectangular rods inscribed with numbers, which became known as Napier's bones (**38**). Each rod is engraved with a table of multiples of a particular digit, the 10s and units separated by an oblique line. To obtain a product of 123 x 4, the rods 1, 2 and 3 are put alongside each other, and the result is read off by adding the numbers in the fourth row: |⅜| |⅞| |⅛|. The adding is done along the diagonal. The number of rods in a set is basically 10, plus extras representing squares and cubes. The rods are usually made of boxwood or ivory, and are often contained in a box or case. Examples can still be found, as the bones were in use until the end of the 18th century. Napier's bones are sometimes found associated with an abacus.

Slide Rule

Another calculating instrument is the slide rule. It is based on logarithms, and its origin is in the scale devised by Edmund Gunter in about 1607, and published in 1623. The Gunter scale is composed of two scales of the logarithms from one to 10 placed end to end, and used with a pair of dividers to connect proportions between numbers on the two scales, so allowing multiplication and division

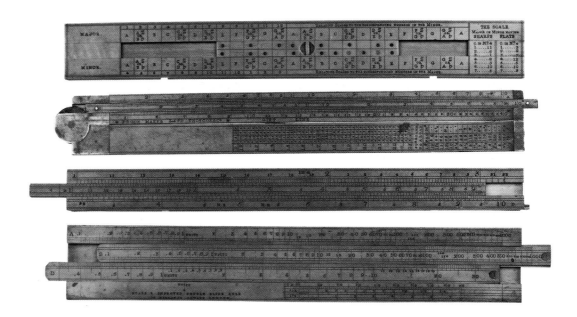

Four English boxwood slide rules. From top: musician's, late 19th century; carpenter's, signed:
F.B. COX MAKER *LATE* THOS COX & CO, *c.* 1830; excise officer's with four slides, one in each side, signed:
DOLLOND LONDON, *c.* 1830; arithmetical, signed:
HOARE'S IMPROVED DOUBLE SLIDE RULE 12 BILLITER SQUARE LONDON, 1867. (MHSO)

to be performed. By 1621, William Oughtred had invented the slide rule, by putting two Gunter scales side by side. There was a controversy between Oughtred and his pupil Richard Delamain over the invention of the circular slide rules which Delamain published under the name of *Mathematical Rings* in 1630, and Oughtred published as *Circles of Proportion* in 1632. Thomas Browne, a joiner, invented a third form – the helical slide rule – which was also produced in the 1620s, with the lines drawn on the surface of a cylinder.

The two most common types of slide rule in the 18th and 19th centuries were the Coggeshall and the Everard. The former, typically used by carpenters, is named after Henry Coggeshall, a mathematician who invented it in 1677. It is a 2-foot jointed rule, made of boxwood with a brass hinge and end-caps, with a brass slider in one arm. The slider bears a Gunter-type logarithm scale; on one side is a conventional double 1-10 scale, while on the other is a broken Gunter scale from 4 to 40, which is called the girt line, for measuring the volume of timber. On the back is the 24-inch rule, and on the back of the brass slider is a 12-inch rule, so that the whole thing can measure a yard. Also marked on the rule is a table which gives the price in pounds, shillings and pence of various units of timber. Along the edge of the rule is a scale dividing the foot into 100 parts, a foretaste of metrication.

The excise officer's gauge for assessing the duty on wines, spirits and ale was almost square in section, with sliders in all four sides.

This was the Everard slide rule, invented by Thomas Everard, a gauger in the Excise at Southampton in the 1680s. It is a foot long, and an inch square, with inset Gunter's scales, and is marked with standard points important for excise use, including WG – wine gallon; AG – ale gallon; and MB – malt bushel.

A large cylindrical slide rule was patented in 1881 by Edwin Thacher, and made in New York; but the handiest cylindrical rule, equivalent to 8 times the length of a normal slide rule, and giving the possibility of reading to 3 or 4 places of decimals, was the invention of Professor G. Fuller at the turn of the century. This was in general use in the early 20th century. The Fuller slide rule was sold in a box, and was fixed by its handle to a bracket at one end of the box when in use.

The traditional, rectangular slide rule is made of boxwood, with brass fittings. Circular models were made of brass, while cylindrical ones were sometimes printed on paper from engraved plates, and mounted on wood. The invention in the 1890s of ivorine, a trade name for a white synthetic material, enabled the development of more accurate engraved scales for all three types of slide rule.

Calculating Machines

Arithmetical machines, which used gearing to speed up calculation, began to be devised at the beginning of the 17th century. There were five types of machine:

Addition machine. The first was devised by Blaise Pascal, a Frenchman, who, after many attempts, produced his definitive model in 1645, and had it manufactured for sale. Samuel Morland, who became Master of Mechanicks to King Charles II in 1681, produced his version of an adding machine in the 1660s.

Addition machine for multiplication. Made by Gottfried Leibnitz in 1672.

Multiplication machine. Invented by Léon Bollés in 1888.

Difference machine. Two separate models were invented, devised by Johann Helfrich von Muller in 1786, and by Charles Babbage in 1822.

Analytical machine. A prototype was invented by Babbage in 1834.

Of these machines, the first to be commercially manufactured in 1820 was by Chevalier Charles Xavier Thomas, of Colmar in Alsace, who had a workshop in Paris. He made his 'arithmometer', which consisted of a long wooden box, with numbers appearing in brass-framed 'windows', and a handle. Numbers up to 999,999 may be set for addition, multiplication, subtraction and division. His machines bear his name. An imitation, which was circular in shape, was patented by Joseph Edmondson of Halifax in Yorkshire. A later patent was taken out by S. Tate in 1903, in the rectangular form. One example in a British museum was used in 1910 by the Royal Automobile Club for speed trials. The arithmetical machine was perfected by Curta of Liechtenstein; it consisted of a little cylinder which could be held in one hand, with a handle at the top. Used by scientists and engineers in the 1950s, it was superseded only by the electronic calculator.

The Culpeper design of compound microscope lasted for over a century. This all-brass version is signed with a trade card in the mahogany case: BANKS, *No 441 Strand, London; c.* 1810. Robert Banks was working from 1796 to 1831. His name is sometimes found as Bancks. (RMS)

OPTICAL INSTRUMENTS

T HE Greeks understood the properties of mirrors and burning glasses and they studied geometrical optics; but the first person to make a study of lenses was the Islamic philosopher, Alhazan, at the beginning of the 11th century. His work influenced Roger Bacon, a 13th-century Franciscan, who made a great reputation as a university teacher at Oxford. Bacon has been falsely credited with the invention of spectacles, but he was concerned with the metaphysics of light, not with practical technology. Eyeglasses (the better name for them) were almost certainly first produced in Pisa, Italy, in about 1286, and were in general use throughout Europe during the Middle Ages. Yet no one seems to have had the idea of placing one lens in front of another to make a telescope or microscope, the difference between the two instruments, in these early days, being dependent upon the distance separating the two lenses.

Although Galileo is popularly supposed to have invented the telescope, the inventor of both the compound microscope and the refracting telescope (as opposed to the reflecting telescope which used mirrors of polished metal instead of lenses) was Sacharias

Janssen, spectacle maker of Middelburg in the Netherlands, in 1608. The early telescope and compound microscope were not of much use by today's standards, though what they revealed appeared marvellous at the time. The glass was poor in quality, and chromatic and spherical aberrations blurred the image.

Chromatic aberration is caused by the unequal refraction of light rays of different colour, which means that the blue and red ends of the spectrum come to a focus at different points, resulting in a coloured edge to the microscopic image. With a spherical lens (non-spherical lenses have only been produced in very recent times), the focal point of those rays that pass through near the edge of the lens is closer to the lens than that of the central rays. As rays from all parts of the object pass through all parts of the lens, the entire image is unsharp.

It was because of these deficiencies that one of the most successful of the early microscopists, a Dutch merchant named Antoni van Leeuwenhoek, used from the 1670s not a compound but a simple microscope, consisting of a tiny bead lens, held up to the eye in a metal plate. With infinite skill, Van Leeuwenhoek achieved significantly higher resolution than was possible with a typical compound microscope even in 1800, a century or more later than he was working.

The compound microscope, nevertheless, had already made possible sufficiently striking observations. In 1660, Marcello Malpighi observed the blood capillaries, thus providing the crucial evidence to confirm William Harvey's theory of the circulation of the blood. Following Malpighi's observations, virtually every microscope sold for the next 200 years had as an accessory the fish-plate or frog-plate, to hold a specimen so that the circulation of the blood could be observed. Other subjects popular with 17th-century microscopists are all described and illustrated in Robert Hooke's *Micrographia*, published in 1665 by the man who was Curator of Experiments to the Royal Society of London, and a former assistant to Robert Boyle.

During the early years of the 18th century, considerable advances were made in the design of the microscope stand, using brass rather than the earlier pasteboard and wood, and giving the instrument greater stability and finer adjustment.

But the first really important optical improvement was the work of John Dollond, the son of a Huguenot weaver, whose hobby was geometry. In middle life, he joined his son Peter in his instrument-making business, and achieved the remarkable feat of correcting chromatic aberration in the telescope by using a combination of crown glass and flint glass for the lenses. The Dollonds marketed lens combinations of this type from 1758, for use in telescopes. The much smaller lenses needed to bring about the same correction for the compound microscope presented greater tech-

nical problems, and were not produced until the end of the century, by an Amsterdam instrument maker, Harmanus van Deijl.

In the course of the 1820s, the problem of making objectives for the microscope that were not only achromatic but also free from spherical aberration was tackled in a variety of ways, including the further development of the reflecting microscope, and the use of lenses made of gem stones. In 1830, however, came the publication, in the *Philosophical Transactions of the Royal Society*, of a paper by Joseph Jackson Lister, in which he described his method of using two achromatic lens combinations at a certain distance apart to eliminate spherical aberration from the image. Lister, a wine merchant, and the father of the great medical pioneer, Lord Lister, crossed the final technical frontier in the development of the optical microscope, which became, during the Victorian period, a vital tool in medical and other scientific research.

Until John Dollond's work on achromatic lens combinations for the refracting telescope bore fruit in the latter half of the 18th century, the best astronomical telescopes were of the reflecting type. The first reflecting telescope was made in 1668, to his own design and with his own hands, by Isaac Newton.

In the 18th century, the London instrument-making trade included many fine telescope makers, of whom the most notable was the Scotsman, James Short. His skill at founding and polishing by hand the pairs of metal mirrors which provided the optics of his telescopes was such that they were bought by most of the observatories in Europe and the New World. Short's achievement was built upon and extended by William Herschel at the end of the 18th century. Large reflecting telescopes were in use in observatories throughout the 19th century, and, with ever bigger mirrors, are still employed today.

The early refracting telescopes suffered from the poor quality of glass used for their lenses, and from aberrations in the image. The extremely long tube required to achieve a focus made them unwieldy and impractical. But from the mid-18th century onward, refracting telescopes were used for nautical and military purposes, and, as the optics improved, for astronomical work as well.

Microscopes

The two basic types of microscope are simple and compound. The simple microscope consists of one lens; or, in some cases, of one lens composed of two or three elements, which simply looks like a single, rather thick lens (**41**). The compound microscope has at least two, usually three or more lenses, which have to be held at a fair distance from each other by being mounted in a rigid tube, made of pasteboard, ivory or, most commonly, brass (**40**). The typical 18th-century compound microscope has a small objective lens (there are probably five or six alternatives to choose from in the accompanying kit of accessories), a field lens of 1–3 inches (2.5–7.5 cm) in diameter, placed in the middle of the tube, and an eye lens (closest to the eye), which is bigger than the objective lens, but smaller than the field lens. The sizes and positions of the lenses vary; Benjamin Martin, for example, often set five lenses in his microscopes. In the 18th century almost all these lenses would have been double convex. In the 19th, when

serious design was for the first time applied to the optical system of the microscope, lens elements became much more complex. The multiple objectives of Victorian microscopes should not be dismounted for examination, except by an expert.

17th-century Microscopes

Few microscopes from this period remain, even in museums. Compound microscopes were made from wood, pasteboard and vellum, which was often decorated with gold tooling. Leading makers were Eustachio Divini and Giuseppe Campani of Rome, John Marshall and John Yarwell of London. The Italian microscopes were small in size, and of the cylinder type, designed to be held up to the light; a tripod stand could be incorporated. The London-made instruments were generally larger, with a gold-tooled leather or rayskin tube, and could be either of the tripod or side-pillar design, the latter being particularly associated with John Marshall (**40**). Because of the comparatively fragile materials and construction of 17th-century microscopes, the tubes have sometimes survived longer than the stands, and can be found mounted on much later stands.

The simple microscope invented and used by Antoni van Leeuwenhoek in the late 17th century consisted of a tiny bead lens set in a metal plate, with a pin mount to hold the specimen. There are 10 genuine Leeuwenhoek microscopes of this form known to exist, and all of these are in collections. It should be noted that a considerable number of imitations were produced, and these, too, are sometimes found in collections.

Screw-barrel Microscope

At the end of the 17th century, a Dutchman, Nicolaas Hartsoeker, invented the screw-barrel construction for the simple microscope, and the design was introduced into England by James Wilson in 1700. Bearing his name, microscopes of this type became popular after it was described in Harris's *Lexicon Technicum* (1704). Edmund Culpeper was a maker whose name also became associated with this design of microscope. In the early 18th century, many Wilson screw-barrel microscopes were produced, made of ivory or brass, and generally contained in a black fishskin or leather box. Focusing the slides is by means of a wide-threaded cylinder which screws in or out of the main barrel; the optical tube may be fitted with a side handle, or may mount on a brass pillar with a tripod foot. This type of instrument could be converted into a compound microscope by the addition of an ivory tube containing a field lens and an eye lens. Such a model was developed in 1746 by George Adams Senior, who changed the screw barrel for a more serviceable wheel of objective lenses, so that a choice of eight could very simply be brought into the optic axis (**42**).

Culpeper-type Microscope

This type of compound microscope was first produced in the 1720s (**43**). The optical tube, made of wood, pasteboard and leather, fits into a cylindrical support covered in rayskin (often called shagreen). This is held by three legs, rising from a circular wooden base, with a simple brass stage between the legs. These microscopes were originally supplied with a wooden case of pyramidal shape, made from oak or, later, mahogany. Those retailed by Culpeper himself have his trade card stuck at the back of the case, showing his sign of the crossed daggers. Later models in the same form have a wooden box foot with a drawer to hold accessories. This pattern continued until the mid-19th century, but leather, pasteboard and rayskin were gradually superseded by brass, until, by 1780, the whole instrument, with the exception of the box foot, was made of brass. Many examples of the later form of the Culpeper-type microscope exist, but the majority are unsigned. Names that may be found, however, are Adams (**43**), Nairne, Nairne & Blunt and W. & S. Jones. This form of microscope was imitated on the Continent, and some Dutch and French unsigned examples are to be found.

Cuff-type Microscope

Another common and important type of compound microscope was that first made by John Cuff, to the design of Henry Baker, in 1743. The instrument is similar to Marshall's design from the 17th century, in that the body tube was supported on a bracket fixed to a side pillar. This pillar is fitted with a well-made focusing arrangement, which allows for the first time some exactitude in adjusting the microscope for use. The instrument is mounted on a mahogany box foot and was originally contained in a mahogany pyramidal case. Cuff-type microscopes may bear the signatures of Dollond, Gilbert and Martin. In Paris, Claude Passemant adopted Cuff's design and his workshop had a considerable production in the mid-18th century.

The Cuff-type compound microscope was a considerable advance over earlier instruments. This is signed: J. Cuff, Londini Jnvt & Fecit; c. 1750. The first of these microscopes was made in 1743 for Henry Baker for the study of crystals, and this work gained him the Copley Gold Medal of the Royal Society. (RMS)

Benjamin Martin's Designs

By the 1780s, the range of microscopes on the market included, as well as the Culpeper-type and the Cuff-type, a third model, designed by Benjamin Martin. This had a tripod foot which folded flat, and a compass joint at the base of the pillar which enabled the pillar to be moved to an angle away from the vertical, to facilitate use. This design was sold by Martin himself and by Adams and W. & S. Jones. Variations on this model were developed and modified into the 19th century.

Another design of microscope which continued into the 20th century was Benjamin Martin's drum. This consisted of a cylindrical tube with a portion cut away near the base to allow light to fall on the substage mirror. It was rather like a very compact Culpeper-type microscope. The model was cheap to buy and therefore popular with amateur microscopists in the period 1820–50. It was available as a cheap toy well into the 20th century. Also popular was Martin's solar microscope for projecting an image on to a screen (**44**).

Naturalists' Microscopes

In the latter half of the 18th century, the study of botany became extremely popular, and several portable versions of the microscope were designed for naturalists. One of the first of these to achieve wide sales was the Withering botanic microscope, invented in 1792 by the famous botanist, Dr William Withering. The pillar, holding a simple lens and a wooden stage, is fixed to the inside of the hinged lid of a small box. When the box is opened, the microscope automatically positions itself ready for use.

Another version, often known as the Cary-type (it was retailed by William Cary among others), is properly called the Gould-type. Charles Gould designed a small compound microscope in 1820 to satisfy the wave of interest in natural history. The parts pack away into a pocket-sized mahogany box, the set consisting of an optical tube with a conical nose-piece, mounted on a slim pillar with rack focusing on the stage.

Nuremberg Microscopes

The town of Nuremberg in Bavaria has long been famous for its superb craftsmen, and also became known as a centre for the manufacture and distribution of simple wooden toys, some of which were made in the Black Forest and the Austrian Tirol. Among these were microscopes made from soft wood, with draw-tubes of card and patterned paper, light in weight and inexpensive. Two designs were popular, the Culpeper-type on three legs, and the drum-type, rectangular rather than round in section. Most of these microscopes were distributed by two traders, whose initials are burnt into the base of the instrument with a hot iron: IM and JFF; others found include CHL and ICR. Contrary to what has often been written, the majority of these instruments date from the first half of the 19th century, although a few may have been made at the end of the 18th century.

Mid-Victorian Compound Microscope

Following the production of achromatic objective lenses, and the work of J. J. Lister in overcoming spherical aberration, the optical microscope moved steadily towards the achievement of the highest resolution of which it was capable. It became an important scientific instrument, with makers such as Andrew Ross, James Smith and Hugh Powell improving the rigidity of the optical tube, and the coarse and fine focusing. The firms who were world-famous for top quality instruments were Ross, Powell & Lealand and Smith & Beck; but there were a great many other makers supplying the amateur market, and many microscopes of this period carry the names of optical firms which merely retailed instruments. From 1860, binocular tubes were

An example of Powell & Lealand's
'No. 3 – Smaller Microscope Stand', signed and dated:
Powell & Lealand, 4 Seymour Place, Euston Square,
London. 1850. First produced in 1843, the limb is
underslung on trunnions, and has a pair of struts linking
the body-tube to the arm above the top of the limb.
There are five objectives, ranging from 2-inch (5.8-cm)
to ⅛-inch (3-mm) focal length. Below the mechanical
stage is the centring mechanism for an achromatic
condenser. The old-fashioned long brass fish-plate is
among the accessories. (RMS)

provided for many microscopes, to make
extended use easier and less tiring (**45**); so
Victorian microscopes of this type are to be
found. Microscopes were also sold in large
mahogany chests, beautifully fitted with up to
100 accessories, many of which would never
have been used. In Europe, the German optical instrument trade rose to prominence with
the new designs of Utschneider and
Fraunhofer of Munich around 1810, Plössl in
Vienna, Schiek of Berlin, and later those of

Carl Zeiss of Jena, and Ernst Leitz of Wetzlar. In France, leading makers were the Chevalier family and the firm of Nachet. With the greatly increased production of the late 19th and early 20th centuries, new models were produced every few years by leading firms (**46**), and catalogues of the companies are the best source of information on dating and identifying their microscopes. Some old catalogues, for example.those of the company Nachet, have been reprinted.

An advertisement in a book published in 1865 lists the following British and European microscope makers. British: Baker, London; Bryson, Edinburgh; Charles Collins, London; H. & W. Crouch, London; Dancer, Manchester; Field, Birmingham; S. Highley, London; King, Bristol; W. Ladd, London; Murray & Heath, London; Parkes & Son, Birmingham; Pillischer, London; Powell & Lealand, London; Ross, London; Salmon, London; Smith, Beck & Beck, London. Continental: Amici, Modena: Beneche, Berlin; Brunner, Paris; Chevalier, Paris; Hartnack & Oberhauser, Paris; Hasert, Eisenach; Kellner, Wetzlar; G. & S. Merz, Munich; A. Mirand, Paris; Nachet, Paris; Plössl, Vienna; Schroder, Hamburg; F. W. Schiek, Berlin; Zeiss, Jena.

Telescopes: Refractors

In a refracting telescope, all the optical parts are lenses, and the name derives from the fact that light is refracted when it enters glass (**47**). This is the earliest type of telescope, and it developed in three different forms. The first is the astronomical telescope which has two lenses, both of which converge the light, and are known as positive lenses, producing the image upside down (which presents no problems for observations of the skies). This type of refractor is very seldom found today. The second form is that for terrestrial use, having a three-lens erecting eyepiece system in addition to the objective lens. All these lenses are positive, and the resulting image is upright. Examples can still be found of this type of telescope from the late 17th and early 18th centuries, made from pasteboard covered in dyed vellum or fishskin.

In the mid-18th century, this form of telescope had long wooden tubes with brass lens mounts. From 1800, improved techniques for constructing brass tubing made telescopes far more compact, with shorter sections of tube that nested or 'telescoped' into each other. The brass body tube was often painted, or sometimes covered with wood or leather, and, in the case of naval telescopes, plaited rope. Some were lacquered and elaborately decorated. The third type of refractor consists of an objective which is a converging positive lens, with a diverging, or negative, eye lens. These telescopes are known as the Galilean type, because Galileo popularised the arrangement from 1610. The advantage is cheapness, because only two lenses need to be figured, and the image is upright. The disadvantage is that magnification is not great. This arrangement was, in the 18th and 19th centuries, chiefly restricted to opera and field glasses. Opera glasses of the late 18th century and the Regency period are often remarkably opulent, made of ivory or ebony, encrusted with pearls or ormolu decoration. In the late Victorian period, many binocular field glasses were produced to the Galilean pattern, as are present-day opera glasses.

Makers

From 1690 to 1710, the most popular makers of telescopes in Europe were Eustachio Divini and Giuseppe Campani of Rome, and John Yarwell and John Marshall of London. In the 18th century, the names of Dollond, Martin and Ramsden appear most frequently on British-made instruments (**49**); in France, Passemant was the leading name; in Germany, Georg Brander in Augsburg made telescopes and microscopes, along with surveying instruments; and in Holland, Harmanus van Deijl specialised in optical instruments.

During the early 19th century, three to four draw-tube pasteboard telescopes, covered with decorated paper, were made in Italy; the name of Leonardo Semitecolo appears on several of them. In the 19th century, a great many names of retailers of telescopes are found. Among the best-known firms of the mid-century are Tulley, Varley & Son, Salmon, Wray, Dixey, Watkins & Hill, Harris & Son, Dollond and Negretti & Zambra. One of the most prolific of French telescope-makers of the Napoleonic period was Lerebours, who supplied the navy, the Bureau des Longitudes, and many surveyors. Other leading French makers were Buron and Lebrun; among those in Germany were Kinzelbach of Wurttemberg, Busch of Prussia and Carl Zeiss of Jena (**48**).

Telescopes: Reflectors

The common reflecting telescope produces an erect image, and was the 18th century's most popular telescope because, using metal mirrors instead of glass lenses, it did not suffer from chromatic aberration. Also, it was considerably shorter in length than the terrestrial refracting telescope, and could be made with a larger aperture, to gather more light. Again, there are three types of reflecting telescope. The Gregorian reflector, named after James Gregory (1638–75) the Scottish mathematician, has a concave objective mirror and a concave secondary mirror, which reflects the light from the objective through a small hole in the middle of the objective mirror into the eyepiece. A similar-looking reflector is known as the Cassegrainian type, where the secondary mirror is convex and the image is inverted. Not many of these were made, but they are occasionally found today. With both these forms, the telescope is put to the eye and pointed straight at the object to be viewed.

With the other type, the Newtonian reflector, the observer stands at right angles to the line of view. This is because the objective mirror collects light and reflects it on to a small plane mirror set at 45° to the axis of the telescope. The eyepiece is, therefore, at right angles to the tube, and at the top end of the telescope. The most popular form of Newtonian telescope was that produced at the end of the 18th century by William Herschel for astronomical observations. The smallest Herschel telescope has a metal mirror of 6¼ inches (15.5 cm) diameter, and has a focal length of 7 feet (2.1 m). The telescope is held in a large mahogany frame. These instruments were made by Herschel for sale, and cost 100 guineas (£105).

Makers

The finest reflecting telescopes of the 18th century were made in London, where all the leading makers advertised as a matter of course in several languages. The reflectors of James Short found their way into most national observatories all over the world. Other leading names are those of Mann & Ayscough, Adams, Nairne and Dollond. In Holland, a leading manufacturer was Jan van der Bildt of Franeker, who manufactured Gregorian telescopes.

Mirrors, Lenses and Persistence of Vision Devices

Mirrors

Sight is the prime sense of man, and optics is the main provider of illusions. Sight could therefore transmit effects that were thought to be magical. This is particularly evident in mirrors with curved surfaces. Arrangements of concave mirrors can give an illusion of an object being within reach. Anamorphoscopes make use of a rectifying mirror, either conical, cylindrical or pyramidal, to make normal a wildly distorted drawing, which can only be seen properly in the mirror. Concave and convex mirrors, mounted back to back, serve to illustrate the difference in the reflected images. Convex mirrors in dark glass were employed as a painting aid by Claude Lorrain, and are named after him.

Camera Obscura

This is one of the oldest of optical tricks, whereby, in a darkened room, a pinhole in a shutter throws an inverted image of the outside scene on the opposite wall. A lens in the hole and a mirror were later used to rectify the image. The portable version consists of a rectangular box, made of oak in the early part of the 18th century, but later of mahogany, with a lens at one end, in an adjustable tube, and a mirror set at a 45° angle to reflect light on to a horizontal ground glass screen. There are 'straight-through' versions without the mirror. Another camera obscura accessory which can still be found is the scioptic ball, used for fitting into a window shutter. This is about the size of the wooden ball used in the game of bowls; inside the hollow ball, usually made of the very hard wood lignum vitae, are two lenses. The ball is set in a rectangular mount of mahogany. The ball mounting allows considerable movement to pick up a variety of scenes.

Camera Lucida

This invention of William Wollaston in 1806 employed a prism that would enable a scenic view and the artist's drawing paper to be seen simultaneously. It could be adapted for use with the microscope, to aid the drawing of microscopic objects.

Magic Lantern

The magic lantern, using light projected through a transparent slide and enlarged by a lens to produce an image on a screen, has existed since the mid-17th century. Examples from the Victorian period can be found, but the majority of magic lanterns on sale date from the early years of the 20th century (**50**). The light source can be either an oil lamp or gas, often provided by a portable gas cylinder; later, magic lanterns had a fitting for an electric bulb. Many slides for use with magic lanterns can be found, the most attractive being made of hand-painted glass (**51**).

Zograscope and Stereoscope

Optical illusions appear to have attracted elaborate Greek names. The zograscope of the mid-18th century consisted of a lens and mirror for looking at coloured engravings, making them appear three-dimensional. It continued in use until at least 1870, but was joined in the 1830s by the stereoscope (**52**). Stereo-viewers could consist of boxes, or even large cabinets, often made of walnut, fitted with pairs of viewing lenses and transparent glass slides. By 1900, stereo cards were being mass-produced, on which pairs of photographs were pasted. These were observed through simple hand-held stereo-viewers, consisting of a pair of lenses and a wire support for the photographic card. Some that survive have a hood of chased aluminium, and were made in the United States in the early years of the 20th century. The cards show views from all over the world, Rome, Chicago and the Holy Land being popular subjects.

Kaleidoscope

This consists of a tube in which two mirrors at 60° to each other were used to make symmetrical patterns from a random collection of chippings of coloured glass. It was patented by David Brewster in 1817, and continues as a popular child's toy today.

Persistence of vision devices

The simple thaumatrope of 1825, a cardboard disc spun between the fingers with, for example, a drawing of a bird one side and its cage on the other, engaged the persistence of vision effect in the eye and marked the beginning of the road that led to the cinema. Other milestones on this road were the phenakistoscope of 1832 and the zoetrope of 1860. In these a series of drawings was viewed through slits in a rotating disc and a rotating drum respectively, so giving the impression of movement. The praxinoscope of 1877 was an elaboration of the zoetrope which made use of mirrors. It could be arranged with a frame to give the impression of a theatrical performance. Muybridge was the first to analyse the movement of animals when, in 1880, he photographed a horse galloping, with a battery of forty cameras, The prints from these photographs were arranged on a wheel, called a zoogyroscope which, when spun in a projector, re-created the movement. Toy versions of these wheels were made during the 1880s for projection in the magic lantern. After the turn of the century, when cine cameras had been produced, sets of prints were reproduced in small books whose pages were 'flipped' by the thumb, so giving an impression of movement. These flip books were cheap and had a considerable vogue.

Cameras

The origins of photography lie in two sciences, optics and chemistry. The optical forerunner of the camera, from which it gets its name, was the camera obscura. The chemistry of photography began in the 18th century, when scientists were working on the photochemistry of silver salts. At the end of the century it was established that light at the violet end of the spectrum is especially effective in darkening silver chloride. The first fixed photographs, using bitumen of Judaea as the sensitive substance, were made in France by Nicéphore Niepce between 1825 and 1827. Twelve years later, his colleague, Louis Daguerre, published his invention of the first successful photographic process. Almost at the same time, an Englishman, William Henry Fox Talbot, perfected his different process. Thus photographs from the 1840s are almost without exception either the metal daguerreotypes, or the sepia, paper calotypes of the English process. In 1851, Frederick Scott Archer invented the wet collodion process, and for the next 40 years wet plate photography was practised.

The next big change was slow to be adopted; although gelatin dry plates were invented in 1871, it was not until the late 1880s that they were sufficiently perfected to bring in the new era of the hand-held camera and the familiar snapshot.

The first photographic cameras were the direct descendants of the box camera obscura, which threw an image on a ground glass screen, and was used by amateur artists. Early cameras usually consisted of two boxes, one sliding within the other for focusing. To make them easier to transport, the boxes were often made to fold down. A camera of this type was used from 1856 by Lewis Carroll, author of *Alice in Wonderland*, who was a skilled photographer. In 'Hiawatha's Photographing', he describes the elaborate process of portrait photography, using a camera of 'sliding, folding rosewood'. By the end of the 1850s, the bel-lows camera was becoming popular, but it was still large, used either on a large stand in the studio, or on a tripod for field work. Early cameras were technically very simple, with no shutters because of the very long exposure time, and no separate viewfinder.

The introduction of relatively fast dry plates in the late 1870s meant that cameras made after 1880 are often hand-held, and have a shutter. They were equipped with magazines of plates or cut films, and had a changing mechanism to bring forward each plate or film for exposure. In 1888, roll film was introduced, and metal began to replace wood for camera construction. At the end of the 19th century, there was a vogue for so-called detective cameras, which were small hand cameras made to resemble some common object, such as a watch, parcel or handbag. These were the forerunners of the precision miniature camera. The first small roll-film camera of the box type was produced by Kodak in 1888 and, together with the folding bellows type, made photography into a popular amateur activity.

The final landmarks in the development of photography were the arrival of the first miniature 35 mm camera in 1924, and the invention of polaroid hand cameras and film by Dr Edwin Land in the 1950s.

PHILOSOPHICAL INSTRUMENTS

I N the mid-17th century, the concept of experimentalism, first proposed 50 years earlier by Francis Bacon, began to take firm root. It superseded the theories about the natural world based on Greek thought, which tried to draw deductions from grand hypotheses, and neglected experimental techniques. Experiment rather than argument had priority at the meetings of the Royal Society of London, which was founded in Christopher Wren's room at Gresham College in 1660.

At the same time, at Cambridge and Oxford Universities, and in the Netherlands at the University of Leiden, experimental philosophy was being taught with the use of apparatus. This led to the vogue for lectures illustrated by demonstrations of physical effects as a form of entertainment. From the beginning of the 18th century, the popularity of the lecture demonstration spread from Poland to Portugal, and across to Harvard College in New England in the United States. Lecturers used a lot of apparatus, and many of those who attended bought similar pieces to entertain friends in their homes. So the effect of the lecture demonstration was greatly to stimulate the manufacture of scientific apparatus.

This popular taste for experiment did not die out by the end of the 18th century, although it became somewhat modified. Much of the apparatus used in lectures came gradually to be employed for teaching elementary science in schools. The very same pieces of equipment that Benjamin Martin, a travelling lecturer and later a leading instrument maker, took with him on his tours in the 1750s around the West Country can be found illustrated in catalogues of suppliers of scientific equipment well into the 20th century (**57**). Still a source of popular entertainment, other pieces of demonstration apparatus have reappeared as toys, for example, the 'drinking duck', based on the thermoscope of Galileo.

In the 18th century philosophical instruments were understood to be apparatus used to demonstrate and study mechanics (including working models); magnetism, pneumatics, hydrostatics and hydraulics, electricity, heat, sound and light. They included meteorological instruments as well. Chemistry was dealt with separately, and, for obvious reasons, was generally practised in the laboratory rather than at home. Old chemical equipment is rarely found, because, being of glass or earthenware, it tended to get broken. Portable analytical kits are sometimes found, as are chemical balances, which are dealt with in Weights and Measures, Chapter 7. (Astronomical models are dealt with in Chapter 1, and the most important optical instruments in Chapter 5, although some of the apparatus used to demonstrate the properties of light are described on these pages.)

Of the vast range of apparatus used over two centuries to demonstrate natural phenomena, a selection of the items commonly found in collections is described here. The purpose is to help in the identification, which poses difficulties in collecting this class of objects. Many may have been made in this century, as schools are now turning out their old laboratory cupboards. But these are still worth preserving, for their design almost certainly dates back to the 18th century. The more elegant pieces, made of fine wood and polished brass, date from periods earlier than the Victorian, and command higher prices. The best sources of identification are the catalogues produced from 1900 to the 1930s by English firms of scientific apparatus manufacturers such as J. J. Griffin and Baird & Tatlock.

Mechanics

Various mechanical effects were demonstrated by models, as described below (**57**).

Forces The various effects of different forces acting on moving bodies – for example, a boat crossing a flowing stream has to have its bows pointing upstream in order to travel straight across – were demonstrated by boards or frames with sets of pulleys and weights, known as parallelogram of forces boards.

Gravity Some of the most popular models are the Leaning Tower of Pisa, to show the line of the centre of gravity; a human figure of an acrobat, holding a balancing bar and poised on a pillar to show equilibrium; and a double cone or cylinder so weighted that it appears to roll up a slope.

Inclined Plane Various models serve to show the amount of force required to draw a carriage up slopes of different inclines.

Inertia The centrifugal machine, or 'whirling table', was used to demonstrate all the different effects of force acting on a mass in uniform circular motion. The same effect can be seen today in a spin-dryer, and has been developed as a fairground sideshow, in which the centrifugal force of a revolving drum holds bodies against its walls. The gyroscope, still popular today as a toy, shows the same effect (**62**).

Levers and Pulleys Sets of levers and pulleys were made for purposes of demonstration. The pulley combinations were often mounted in a large wooden frame. Models of winches and jacks, both using the lever principle, were also made.

Fall, Projection and Momentum In the latter half of the 18th century, George Atwood, a mathematician, devised his fall machine, which was a 6-foot (1.8-m) high piece of apparatus, designed to show the laws of motion of bodies uniformly accelerated and retarded (**59**). It included a clock mechanism, a measuring scale and complicated arrangements of pulleys and weights. There were numerous accessories. Atwood fall machines are rare.

Model of an Archimedean screw, used in the Clarendon Laboratory, Oxford, in the mid-19th century. When the helix is turned, a ball placed at the bottom will be raised to the top. (MHSO)

Also popular were boards on which a marble could be rolled to show the parabolic curve of a projectile. Another demonstration shows that a ball takes exactly the same time to fall across the diameter of a circle as it does to fall along a chord of the circle, that is, the shorter distance between two points in the semi-circle (**58**). The effect of momentum and collision was demonstrated by a series of balls suspended side by side at the same height from a framework which can be swung to knock

against one another. This has now become a modern toy.

Mechanical Models. Models of hoists, cranes, mills and pile-drivers and Euclid's three-dimensional theorems were popular, because they showed the practical application of mechanical principles (**39**). The use of a steam jet was known from Greek times, but steam harnessed to produce power was gradually developed only through the 17th and 18th centuries, culminating with the work of James Watt in the 1760s.

Models of the different types of steam engine, bearing the names of their inventors, Branca, Savery, Papin, Newcomen and Watt (**60**), were made for demonstration, and miniature locomotives were also produced during the railway boom.

Magnetism

A natural magnet, or lodestone, is a form of iron oxide, Fe_3O_4, found in various parts of the world but particularly in Siberia. There are ancient references to the magnet in Chinese, Egyptian, Greek and Roman texts. The name comes from the Greek, for magnets were found in the Greek province of Magnesia. The Chinese for magnet translates as 'love-stone' and the French is *l'aimant*.

The earliest application of the magnet was in the compass, whose origin in the West is assigned to Amalfi in Italy, in the 12th century. Once the use of magnetised iron for direction finding was known, magnets became commercially important (**63**). Every seaman needed one to keep his compass needle magnetised. They were also used as philosophical instruments to demonstrate the power of magnetism. There was a demand for very large magnets from Siberia – the Ashmolean Museum in Oxford was presented with one in 1756 that would support 163 lbs (*c.* 74 kg). Russian lodestones are sometimes found in filigree brass cases, inscribed with the weight that the magnet will support.

Pneumatics

Natural philosophers studied the nature and properties of air and gases, and were particularly concerned with creating a vacuum. Otto von Guerike of Magdeburg, in 1654, performed the classic vacuum experiment, using a very primitive air-pump, by evacuating the air from two large iron hemispheres, fitted together. Once this had been done, no amount of force would tear the two halves apart. The so-called Magdeburg hemispheres were still being produced in the 1930s.

The first really practical single-cylinder air-pump was devised by Robert Boyle and Robert Hooke, but it was Hauksbee who made the two-cylinder model that continued in general use in Britain (**55**), although on the Continent, large-bore, single-cylinders were usual, such as those produced by the Van Musschenbroek family around 1700. The air-pump had a range of accessories (**56**). Air-pumps may be found today, although the glass receiver is likely to be a replacement.

Hydrostatics

Strictly speaking, this is the term used for the equilibrium of fluids, and the pressures they exert, while hydrodynamics concerns the motion and flow of fluids, and hydraulics the construction of machines that use fluids. But hydrostatics was the name given to apparatus used for demonstrating all effects involving water. Some of these were very attractive. Diving bells were made, containing a tiny human figure; the Cartesian diver could be made to move up and down in a column of water by pressing a membrane over the top of the container; and the so-called Tantalus cup demonstrated the effect of a syphon, again sometimes using a human figure. There was much interest during the 18th century in the use of water for ornamental purposes in elaborate fountains, and demonstration pieces of brass and glass exist. Another common item was a series of thin glass tubes in a frame to show capillary attraction. Other practical instruments were the hydrometer, or areometer, and the gravimeter, devised to measure the specific gravity of liquids and solids respectively. Hydrometers were used by brewers and excise officers. The Clarke pattern dates from 1730, and the more accurate design of Sikes was introduced by Act of Parliament in 1817. This was a popular excise officer's instrument, and can be found in its box, accompanied by a thermometer, glass flask, sliding computing scale and tables. Hydrometers for specific gravities greater than water, for acids and alkalies, were sold by William Twaddell of Glasgow from about 1800.

Electricity

The word 'electricity' derives from the Greek word for amber, since the Greeks were aware of the attractive quality of rubbed amber. However, it was not until the 18th century that the effects of frictional electricity were studied (**64**). Hauksbee produced the first 'frictional electrical machine', consisting of a round glass globe, revolved by a handle, with a pad of leather pressed against it. In dry conditions, electricity can be generated by rubbing on the surface of the glass. Later versions of the frictional electrical machine were a cylinder of glass (patented by Edward Nairne in 1784 as a 'medical electrical machine') and the plate machine, consisting of discs of glass, spun between two pairs of rubbing cushions. These discs were occasionally made up to 6 feet (1.8 m) in diameter for demonstration machines to produce dramatic effects, but the common type of plate machine had a diameter of 1–1½ feet (30–45 cm). It should not be confused with the later 'Wimshurst induction machine', which has oblongs of metal foil around the edge of the glass discs (**66**).

Following the invention of the frictional electrical machine, the next development was the invention in 1745 by von Kleist in Germany, and Musschenbroek in Leyden, of the Leyden jar, which condensed the electric charge and enabled greater sparks to be produced (**65**). These are glass jars of varying size, covered with foil, with a conductor through the cork in the neck. They can still be found, as can the insulating stools used in demonstrations, which have glass legs.

Some of the equipment used for demonstrating frictional electricity is also occasionally located. This may include carved heads with human hair, intended to be made to stand on end by the electric charge; the 'gamut of bells', which consisted of eight bells on a stand linked by an electric whirl carrying a clapper that, in revolving, struck each bell in turn; and the 'thunder house', intended to show the need for a continuous metallic lightning conductor, in which an explosion, caused by a tiny charge of gunpowder, was set off at the point of discontinuity by an electric charge.

The natural phenomenon known as the aurora borealis was recognized during the late eighteenth century as being produced by an electrical discharge in rarified air. The aurora flask or tube to demonstrate the glow effect in a partial vacuum was a highly popular piece of apparatus for display (**67**, **69**). In the mid-nineteenth century, the German glassmaker Johann Geissler perfected the art of making discharge tubes with a high vacuum, which contained rarified gases, such as hydrogen, nitrogen, oxygen and phosphoric acid. The

Columbia Phonograph Co. advertisement in *The Illustrated London News*, 1904.

technology of Geissler tubes also helped to introduce a new branch of physics, leading to the discovery of cathode rays (**68**).

Current electricity was first worked on in 1790 by Luigi Galvani, professor of anatomy at Bologna. He was followed by Alessandro Volta of Pavia, who, at the beginning of the 19th century, developed a current generating device which became known as Volta's pile, or

the Voltaic pile. Two measuring instruments were fundamental once current electricity had been discovered. The first was the electrometer that measures potential difference, recorded in volts, and depends on attraction or repulsion of charges on plates or wires. Developed into high levels of accuracy by William Thomson, Lord Kelvin, the instrument achieved its greatest sensitivity in the quadrant electrometer of

Edison Bell advertisement
in *The Tatler*, 1906.

1867 (**70**). The galvanometer measures electric current, and acts by running the current through a coil to make a magnetic field. Again, Thomson's work was crucial in perfecting the instrument (**71**).

Heat

One of the earliest of demonstration pieces concerned with heat is 's Gravesande's ball and ring, which shows that metal expands when heated. The metal ball will pass through the ring when cold, but will not do so when heated. Pairs of large, brass parabolic mirrors were used to show that heat could be focused, in the same way as light, from one to the other. The effects of heat were also demonstrated by the thermoscope, devised by Galileo, using a column of water in a spiral glass tube which rose as air in a bulb expanded with heat; and by the pulse glass, consisting of two glass bulbs linked by a tube containing coloured spirit: when one bulb was held in the hand, the warmth caused the spirit to boil and flow into the other bulb. A late 19th-century heat device was Crookes's radiometer, consisting of a glass bulb that contained discs of mica mounted on four pivoted arms. These discs are white on one side, and black on the other. When placed in the sun, or near a light source, the black surfaces, readily absorbing and emitting heat, cause the device to revolve rapidly.

Light

Optics, or the study of light, may be divided into three parts, geometrical, physiological and physical. The first consists in tracing rays from a light source through different media, such as water and glass, from mirrors, and through lenses and prisms. Such studies are ancient: for example, Renaissance architects and artists were aware of perspective. Mirrors, mounted prisms, heliostats (**53**) and other apparatus were used to demonstrate refraction, diffraction, interference and the velocity of light.

The way that the human eye deals with light has been extensively studied, and it was Thomas Young at the turn of the 19th century who proposed a theory of colour vision based on three receptors in the eye, responsive respectively to red, green and violet. This theory was revived by James Clerk Maxwell in 1849, and he used a spinning disk with adjustable sectors of coloured paper to work

out quantitative colour equations. For demonstrations, the colour disks were mounted on frames (**54**). Polarisation, an important discovery in physical optics, was identified by Christiaan Huygens in 1690, and produced two instruments, the polariscope and the polarimeter.

The most important instrument devised as a result of the study of light is the spectroscope. In 1814, Joseph von Fraunhofer (1787–1826) noticed that the Sun's spectrum, dispersed by a glass prism, was crossed by hundreds of fine, dark lines. In the 1850s, Gustav Kirchhoff and Robert Bunsen realised that these lines could be used to determine the chemical composition of the Sun, the stars and many other substances. This process, described as spectrum analysis, created a demand for prism spectroscopes; their basic features are a slit and collimator to produce a parallel beam of light, a prism or grating for dispersing the different wavelengths, and a telescope to observe the dispersed spectrum. In its various forms, the spectroscope has perhaps contributed more to modern science than any other instrument. Principal early makers were John Browning and Adam Hilger in London, Schmidt & Haensch of Berlin, and Jules Duboscq in Paris.

Sound

The study of the physics of sound, sometimes called acoustics, was not sufficiently advanced to take a prominent place in the 18th-century lecture curriculum. Only two demonstrations were included: the alarm-clock under the bell-jar that became inaudible when the air was removed by an air-pump; and the effect of still and moving nodes shown when a glass jar is struck and vibrates.

In the 19th century, the study of acoustics excited much interest, the leading figure in research being Herman von Helmholtz, professor of physiology at the University of Königsberg. It was he who inspired Karl Rudolph Koenig (1832–1901), one of his students, to set up his own business in Paris as an inventor and manufacturer of apparatus for studying sound (**61**). He had an unrivalled reputation for novelty and accuracy. Two of his most characteristic products were tuning forks and dome-shaped brass resonators, used to amplify the note produced by the fork. The importance of this research into sound was that it led eventually to the development of two ubiquitous modern devices, the telephone and the gramophone, which evolved into hi-fi. The early inspirational models for both instruments were by Koenig. Early telephones, developed in the mid-19th century by Charles Wheatstone and the German Johann Reis, are popular collectors' items, as are the gramophones associated with the names of Emile Berliner and Thomas Edison.

METEOROLOGICAL INSTRUMENTS

Thermometer

The thermometer, a descendant of Galileo's thermoscope, is used to measure temperature (**72** right). The problems in making thermometers were to find a substance which would respond uniformly to changes in temperature, and to devise a suitable scale for measurement. Daniel Fahrenheit of Amsterdam, working in the early years of the 18th century, was the first maker to produce thermometers of any accuracy. In 1717 he initiated the use of mercury, which, although technically more difficult to handle, proved much more satisfactory than alcohol, having a more uniform expansion rate and a far wider temperature range. Fahrenheit also endeavoured to produce a satisfactory scale. The scale he made in 1724 had three fixed points: 0° was the freezing point of a mixture of ammonium chloride and snow, 32° was the freezing point of water, and 96° the mouth temperature of a healthy human being. This Fahrenheit scale was used on the majority of English thermometers of the 18th century.

Many other scales were devised at about the same time, but only two are important, those of the French physicist, René de Réaumur, and Anders Celsius, a Swedish astronomer; these scales are generally found on instruments of Continental manufacture. The Réaumur scale had only one fixed point, the freezing point of water 0°, rising to 80° (the boiling point of water). Réaumur used alcohol in his thermometers. The Celsius scale is remarkable in that it was the first to consist of 100 degrees. It had the boiling point of water at 0° and freezing at 100°. The year after it was produced in 1742, the Celsius scale was inverted by Cristin of Lyons, into what has been known successively as the Lyonnais scale and the centigrade scale. Today we are reverting to the name Celsius.

There are thermometers made for a huge variety of specialised purposes, from cooking to mining. Early thermometers usually have scale plates of brass, often silvered, and cases of mahogany. More modern thermometers

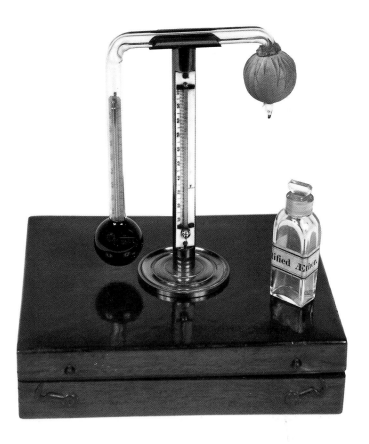

Signed: *J. NEWMAN LONDON* (c. 1830), this dew-point hygrometer was invented in 1820 by J. F. Daniell, professor of chemistry at King's College, London, from 1831. Ether is poured over the muslin-covered bulb, cooling the fluid in the tube, so that atmospheric water vapour condenses on the black bulb (black to aid visibility of the drops). The difference in temperature between the thermometer inside the black bulb and the one outside gives the relative humidity from printed tables. (WMHS)

have ivory or boxwood plates and softwood cases. The maximum and minimum thermometer was devised by James Six of Colchester at the end of the 18th century. It uses an alcohol thermometer to propel a col-umn of mercury, which, in turn, pushes steel indices into place to record maximum and minimum temperatures.

Hygrometer

The hygrometer is intended to demonstrate the humidity of the air. In the 18th century, this was done by a method first devised in 1663 by Robert Hooke, using the beard of a wild oat. The beard is a tiny spiral which unwinds as the tail becomes damp, and this effect is recorded on a dial by attaching an indicator of straw to the oat-beard. This type of hygrometer is often found associated with

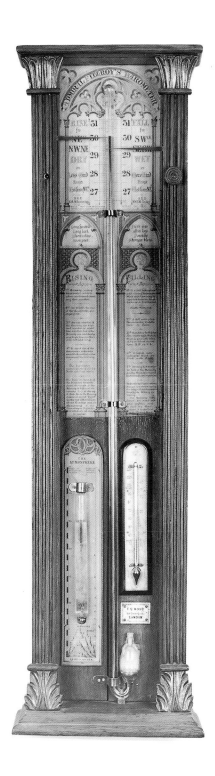

barometers after 1760. Occasionally, cat-gut is used instead of the oat-beard. In 1820, a more scientific method of measuring humidity was devised by J. F. Daniell. His hygrometer consisted of a U-shaped glass tube with bulbs at each end, one painted black, the other covered with silk. The tube contains ether, and there are thermometers both inside the tube and mounted on the pillar supporting the tube. Ether dropped on the silk surrounding one of the bulbs evaporates, and the consequent cooling causes dew to form on the surface of the other bulb. The temperature at which this occurs establishes the 'dew-point'. Later versions of the hygrometer were all developments of Daniell's principle.

Barometer

The barometer is designed to measure the pressure or weight of the air (**72** left and centre, **73**). At sea level this is 15 psi (1 kg/cm^2), and this has significance in the scientific study of the air and gases; but the popularity of the barometer stems from the 17th-century discovery of the connection between alterations in the pressure of the air and changes in the weather. In its simplest form, a barometer consists of a glass tube under 3 feet (1 m) in

Inscribed: Admiral Fitzroy's Barometer. This form is named after Admiral Robert Fitzroy who reorganised the British Meteorological Office from 1854. It also bears a retailer's plaque: E.G. WOOD, 74 CHEAPSIDE, LONDON; late 19th century. This type became popular as an indicator of the weather, with a mercury barometer, thermometer, and a 'storm glass'. This contains a liquid and crystals, which rise and fall and were thought to foretell the weather. (PC) length, sealed at its upper end. The sealed end of the tube contains a vacuum, below which is a column of mercury. The open end of the tube stands in a cistern filled with mercury. An alternative is a bent or siphon tube, with the shorter leg open. The pressure of the air on the surface of the mercury in the cistern, or the open end of the siphon tube, is recorded by the rise and fall of the mercury at the top of the tube. A later development was to bend the tube at an angle of just over 90°, so that the movement of the mercury was magnified and easier to measure.

Most 18th-century barometers are of the stick type, whereby the tube is held in a long, narrow mahogany case, frequently featuring an ornamental top. Others are of the bent tube type, some with a mirror set within the angle, in a picture-frame mounting. A third type, invented in the 1660s, was the wheel barometer, which uses a siphon tube with a float on the mercury and a pulley and weight arrangement to record rises and falls of the mercury on a circular dial. This type, in which the tube is invisible, did not become popular until the early 19th century, when it was the arrangement used in banjo barometers.

Barometers were often associated with thermometers and hygrometers, and many also incorporated a vernier scale for greater accuracy of measurement; they were also associated with clocks.

The first instrument maker to specialise in the manufacture of barometers was John Patrick of London, who was described in the *Lexicon Technicum* of 1704 as the 'Torricellian Operator'. This elegant title commemorates Evangelista Torricelli, to whom the invention of the barometer in 1644 is generally ascribed.

Patrick was sufficiently well known to have been visited in 1710 by the German traveller and diarist, von Uffenbach, who called him 'an optico and weather-glass maker'. Patrick published two pamphlets on meteorological topics, publicising his products.

The most sought-after of antique barometers are probably those made by Daniel Quare, a Quaker clock-maker of London at the beginning of the 18th century. His instruments are made of ivory, silver and fine woods.

Most antique barometers will have had replacement tubes and mercury, because the tubes are easily broken in transit by careless handling of the heavy, liquid metal. Late Victorian and Edwardian stick and banjo barometers can be like the earlier examples, but elaborate, Gothic decoration may indicate their late date.

Barometers dating from the latter half of the 19th century may also have an aneroid barometric mechanism in place of the mercury tube; this was patented in 1845 by a Frenchman, Lucien Vidie. The mechanism consists of a metal bellows, partly evacuated of air, which is very sensitive to changes in air pressure that are recorded by a pointer over a circular dial.

In the first half of the 19th century, an extensive trade in London-made banjo barometers began to be conducted by Italians, who retailed them in hundreds of British towns and cities. These instruments, usually carrying a level, dial and hygrometer (sometimes a thermometer as well), bear the name of an Italian tradesman, incongruously linked to an English place-name. One of the most famous of these firms was Negretti & Zambra, which became a very considerable supplier of

all sorts of scientific instruments, especially meteorological ones, and still exists.

The *baromètre liègeois* is a foreteller of storms and possibly evolved from drinking vessels for birds. It is a pear-shaped, closed glass vessel with a long spout rising from the bottom. When the vessel is half-filled with water, changes in atmospheric pressure will cause the water level in the spout to rise for a storm (low pressure) or lower for fine weather (high pressure). Early 18th century glass- blowers of Liège made these domestic weather indicators, and they have been copied and simplified up to the early 20th century. They are often found in

French barograph for continuously recording barometric pressure by a pen on a cylindrical chart that is turned by clockwork. The pressure is detected by an aneroid barometer. Signed: *Maison de l'Ing^r Chevallier Opt^n, Avizard F^{res} Suc^{rs}, 21 rue Royale, Paris* No. 237; *c.* 1900. (MHSG)

the Netherlands. Being of glass, they are difficult to date, but later models may be etched with the retailer's name.

Letter balance in brass and mahogany on the Roberval principle, which employs upper and lower crossbeams. The postal rates on the platform were in use from 1871 to 1897. (PC)

WEIGHTS AND MEASURES

S YSTEMS of weighing and measuring date from the era of settled, civilised communities, which practised barter and trade. The exchange of goods requires agreement on the amount or mass of the goods offered by each party, in order to assess their equality. There is evidence, much of it pictorial, that men were using weights and scales, capacity measures, and linear measuring rods, from at least 3000 BC. At first, each village or market place had its own standards of weights and measures, which varied according to the goods concerned. The latter custom has persisted into this century with the use of troy weight for precious metals and pharmaceuticals, a different metrological scale from avoirdupois (the commonly known Imperial weight scale). Only under a strong, centralised government were attempts made at standardisation. The Romans, for example, had a single standard weight unit, the *libra*.

Throughout the Middle Ages, a vast and bewildering variety of weights and measures existed, undoubtedly making commerce more difficult. Despite frequent government intervention – e.g. the edicts of Charlemagne and Magna Carta – regional differences persisted within different countries. The first really effective step towards national standardisation was the introduction of the metric-decimal system by the Revolutionary government of France in 1792. During the 19th century, most countries adopted and standardised to either the metric system, or the Imperial (British) system. Now, most countries (excluding the United States of America) have adopted the metric system.

Early civilisations, such as the Egyptian and the pre-medieval Islamic, achieved a remarkable degree of accuracy in their weights. Obviously, the weighing of precious metals, coinage and drugs required strict control to prevent fraud, so most governments introduced methods of certification. This involved the stamping of weighing equipment with city or state symbols. In addition, in medieval Europe, makers of weights put their own stamps, or master-signs, on the weights, as a guarantee of accuracy.

Types of antique scales vary from large steelyards employed for weighing sacks of grain to the delicate balances designed for weighing coins or drugs. Steelyards or extended spring scales were generally used for heavy loads. Domestic scales of the Victorian and Edwardian periods are of the spring type, with brass or tin pans resting on a cross-shaped support. Letter scales, very common throughout the 19th and early 20th centuries, were either balances, or spring scales, sometimes of the self-indicating type. A more accurate balance was required for weighing coins, jewels and chemicals. Chemical balances for laboratory use are of fine workmanship and may be found in their own fitted cases.

BALANCES AND SCALES

Balance

This is the most accurate weighing mechanism and consists of two pans suspended from the ends of a beam, which is supported at its midpoint. One pan is used for the goods that are to be weighed, the other for the weights. Equilibrium of the loads is indicated by an exactly horizontal position of the beam. At first, the equilibrium was judged by eye, but later balances all incorporated an indicator or pointer to show the equilibrium, or lack of it.

Bismar and Steelyard

These two types of scales employ the principle of the lever, with unequal arms for the beam. All scales with unequal arms use a single counterpoise, fitting only one specific scale. There are two methods of weighing with an unequal-arm beam. In the bismar, the counterpoise is fixed to one end of the beam, and the axis of the beam can be shifted quite simply, for example, by a loose loop of cord supporting the beam that can be moved along it.

As soon as the beam is horizontal, showing that equilibrium has been reached, the correct weight can be read off the scale marked on the beam from the position of the cord loop (the fulcrum). Bismars made in Russia, Finland and Sweden in the 18th and 19th centuries are obtainable today.

Steelyards are levers with an immovable axis or fulcrum, which have a pan or hook attached to the shorter arm for holding the load and, suspended from the longer arm, a movable counterpoise. At the place where the counterpoise keeps the beam exactly horizontal, the weight of the object is indicated on a scale, again marked on the beam. The steelyard was extremely common, and still remains in occasional use.

Self-indicating Scales

These have no beam, but consist of one pan or hook for the load on a long arm, with a heavy counterpoise on a short arm. Once the load is in position, the counterpoise swings upwards from its rest position, which is vertically downwards, to balance the load (by the principle of the lever), and an indicator on the long arm points to the inscribed scale, giving the weight of the load.

Spring Scales

Unlike those described above, these do not rely on achieving a balance, but are based on the deformation of elastic material by weight. The elastic material is usually a steel coil, which is compressed or extended by the weight of an attached object, the degree of compression or extension being recorded by a

Chondrometer, or corn scale, English, c. 1840, by Loftus, London. Regulated by an Act of Parliament of 1 January 1826, and used to assay the quality of grain. (Christie's)

pointer on a scale, which has to be calibrated by the maker.

Coin Balances (Money Scales)

These are known from the late 15th century. The earliest type consisted of a hand-held balance with two pans, which fitted into a small box carved from solid wood with individual sockets for the weights. The earliest weights used with these balances were square, but following a Proclamation of Charles I in 1632, English coin weights were decreed by law to be round. Coin balances from the reign of

Charles I are likely to bear his initial, surmounted by a crown, as a verification mark. A boxed balance of the late 17th century may have the maker's trade card stuck into the inside of the box lid. During the 18th century, a number of new designs for coin balances appeared, for there was an increase in numbers of foreign coins circulating all over Europe as a result of expanding overseas trade. Merchants and shopkeepers needed to have the means of checking gold coins both for the weight of gold, and also for base metal content. For the latter reason, a number of hydrostatical balances were produced, with which a coin's specific gravity could be tested by weighing in air and then in water.

One popular coin balance of this period was of the steelyard type, consisting of a slim rule marked with a scale, which could be used on any convenient flat surface, from which hung a clip to hold the coin. This fitted into a pull-apart case.

Following the British recoinage in the 1770s, and the New Standard of 1774, which withdrew from currency coins below a specified weight, another steelyard type of balance was produced, which 'popped up' when the lid of the slim box was opened. A hinged weight could be swung into two positions for weighing the guinea and half-guinea, a great simplification compared with earlier scales. The design is attributed to Englishman Anthony Wilkinson in Lancashire, and various makers' names are found associated with the towns of Kirkby, Ormskirk, Prescot and Birmingham.

In 1817, new coins, the sovereign and half-sovereign, were introduced. To test these coins for wear, and to detect counterfeit coins, the so-called sovereign rocker was invented, a simple little balance in a pull-apart case, which included slot gauges to check the size of the two gold coins. Other designs of coin balance existed, including a spring balance combined with a pen, and a desk-top balance for gold and silver coins. But after 1860, it was increasingly uncommon for the general public to check coins, and the production of coin balances declined.

Alongside the novel designs described above, the traditional two-pan balances continued in use, their date indicated only by the style of the box. Solid wood boxes, with the slots for the balance and weights carved out, are an indication of considerable age, probably dating from the early 18th century; and a fish-skin-covered box, lined with velvet or silk, almost certainly dates from the same century. The commonest type of box was oval, made of japanned metal, first produced in the mid-18th century, but in use for about a hundred years.

Another way of dating a boxed coin balance is by means of the weights, which may be marked for specific coins. It is then necessary only to determine when the coin was in circulation. In early boxes, there may be individual slots for the weights, so it is easy to see if the weights supplied are original; but with later boxes, there is merely a compartment for all the weights, and it is possible that stray weights may have been added. Sometimes there may be a table of coins and their current weight inside the box lid, which will help to verify the weights and to date the box. It is common, however, to find an odd assortment of weights, either added over several generations of use, or by modern owners or dealers.

Chinese 'Opium' Scales

These characteristic scales, which may occasionally be located, have a beam of ivory, with a single brass pan. They work on the steelyard principle, and are contained in a wooden case commonly made in the shape of a banjo or violin, with the beam running up the narrow part, and the pan contained in the wider portion. The weight is of brass, with a rod through it, to which the suspension cord is attached.

These scales are of Chinese origin, and are also called *dotchin*. They are first mentioned in literature in the 17th century, and were used for weighing precious metals, jewels and medicines (with the possible inclusion of opium). They continued to be manufactured well into the 19th century.

Set of British spherical, or 'bomb', weights, c. 1890, ranging from 1/2 oz to 56 lb. Engraved with the name of the Scottish town, Burgh of Govan, whose officials used them. (Christie's)

Apothecaries' Balances

These are distinguished by the glass pans (occasionally ivory) and are generally hand-held. They were intended for use by doctors, who did their own dispensing, and in the home. Typical accompaniments in the box would be a tiny brass shovel and a small glass measuring cup.

Jewellers' Balances

Always mounted on a stand, they are intended for more precise work. The balance beam is very thin and the indicator will be very exact, the pans tiny, and often made of silver.

Chemical Balances

By the end of the 18th century, these required a high degree of precision, and were masterpieces of the instrument maker's craft (**75**). The first attempts to construct an exact bal-ance for scientific use were made at the beginning of the 18th century, and in the following hundred years many of the leading instrument makers of London tried their hand at producing them, including Ramsden and Troughton. Such balances from the 18th and 19th centuries are not very likely to be found outside museums. They are associated with a glass case, normally with drawers below. The case is usually mounted on levelling screws, and the front, and possibly sides, are movable.

Letter Scales

The first general postal service with pre-paid adhesive stamps was introduced into Britain in 1840. From then on letter scales were required to decide the weight and hence the cost of stamps required to transmit the letter. Most Victorian and Edwardian scales are usually of the balance type, made of brass, with a wooden base. The letter rests on a flat rectangular brass plate, the weights on a smaller, circular one. The table of weights and equivalent postage is often inscribed on the rectangular plate. Others are spring balances, with the tariff shown by a pointer. Because of the shape, these were known as 'candlestick' scales.

Grain Scales

Examples exist of a special kind of scales known as a chondrometer, the purpose of which, according to an instruction label in an example made by Watkins & Hill in 1826, was 'to ascertain the quality of all kinds of grain or farina by inspection only'. The scale reads off in pounds per bushel (a capacity measure of 8 gallons) when the brass cylindrical cup is filled with the sample of grain. Early English examples date from the 18th century. Grain scales exist that were made in Germany, Holland, Sweden and France during the 19th century.

Weights

The earliest weights were simply pieces of hard natural stone of convenient size and shape. Egyptian weights, preserved from about 2500 BC, are made of granite. Later, bronze and lead were used, but although metal had the advantage of high specific weight, so that metal weights could be smaller in volume, corrosion was a problem. While the Greeks used lead or bronze for weights, the Romans used stone, often polished serpentine, until the third century AD. Glass, being as durable as stone, was used in Islamic countries in the medieval era. In medieval Europe, iron and brass, or combinations of the two, were employed for weights, with the occasional use of other metals, and sometimes glass or glazed clay. By the 18th century, brass, iron and pewter were the common materials for weights all over Europe. For standard weights that have to keep their accuracy over long periods rock crystal (quartz), bronze and platinum have been used.

Weights come in a variety of shapes. The majority are geometric: spheres, hemispheres, squares, cubes, discs and polygonal forms. But there are also many attractive representational weights, in the form of animals and even human effigies. Pharmaceutical weights were sometimes modelled upon the symbols used in prescriptions, as an aid to accuracy. Another common construction for weights was dictated by convenience: the different denominations were made to fit one inside the other, most simply by using weights in the form of cups. Perhaps the most famous and sought-after of nested weights are those made in Nuremberg in the 17th century. These fitted into a master cup, which was beautifully embellished, and had a carrying handle. Brass nested weights, because of their convenience, were produced well into the 20th century; they often appear older, in fact, than they are, because of the traditional design.

The symbols and inscriptions on weights are a vast and complex subject, and collectors need to consult the tables of the wide variety of weights applicable to different countries at different times. The standard unit in Germany from the 12th to the 19th centuries was the mark, or marc (2 marks = 1 pound = 467 gm), and subdivisions included the loth, the quint and the pfennig. The French pound was the livre (1 livre = 489 gm), and subdivisions included the marc, the once, the gros, the denier and the grain. In England there were two main weight systems: the troy, generally used for precious metals, precious stones and dispensing drugs, and the avoirdupois, the customary commercial system.

In troy weight, 1 penny weight = 24 grains; 1 ounce = 20 penny weights; 1 pound

Postal scale of the 'candlestick' pattern, signed: R.W. WINFIELD BIRMINGHAM; c. 1850. This pattern of spring scale was introduced at the same time as the penny post. Winfield, a brass bedstead maker, registered the design on 13 January, 1840. (PC)

grains; 1 drachm = 3 scruples; 1 ounce = 8 drachms; and 1 pound (troy) = 12 ounces. Note that it is only the grain that is common to all these scales of weights. 1 lb troy = 5,760 grains, and 1 lb avoirdupois = 7,000 grains.

Measures of Length

The earliest units of linear measure were based on approximate body measurements: the fingers and hand, the forearm (cubit), the length of both arms fully extended, measured across the body (fathom), the foot, the stride (giving a mile of 1,000 paces), etc. The inch derived from the width of the thumb, as a twelfth part of the foot, while the cubit measured approximately 18 inches (45 cm) or slightly more; the Egyptian royal cubit was 20.6 inches (52 cm). The Greek and Roman foot measured approximately 12 inches (30 cm), and these continued in use throughout Europe until the medieval period. In England, a statute of 1305 laid down the standard of linear measure, mainly for the purpose of measuring land. According to this, three grains of dry barley make 1 inch; 12 inches make a foot; three feet make an ulna, which later was renamed the yard. Henry VII, in 1497, had the first standard yard made, and this octagonal-section bronze rod still exists in the Science Museum, London. It is 0.037 inches shorter than the Imperial yard used in

= 12 ounces (= 373 gm); and 1 pound avoirdupois in troy measure = 14 ounces 11 penny weights 16 grains.

In avoirdupois weight, 1 dram = 27.3 (nearly) grains; 1 ounce = 16 drams; 1 pound = 16 ounces (= 453 gm); and 1 pound troy in avoirdupois measure = 13 ounces 40 grains.

There was also an apothecaries' scale (superseded in 1864) of which 1 scruple = 20

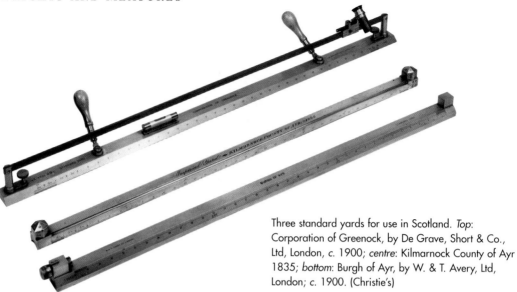

Three standard yards for use in Scotland. *Top*: Corporation of Greenock, by De Grave, Short & Co., Ltd, London, c. 1900; *centre*: Kilmarnock County of Ayr 1835; *bottom*: Burgh of Ayr, by W. & T. Avery, Ltd, London; c. 1900. (Christie's)

the 20th century. Elizabeth I, in 1588, confirmed the yard, but added another linear measure, the ell, which measured 45 inches, and was intended as a cloth measure. The ell continued as a legal measure until 1824.

As with weights, so with linear measures, there was the likelihood of very considerable variations from country to country, and also between different towns. The Scottish ell, for example, measured 37 inches (94 cm), and the Flemish ell 26 – 27½ inches (68–70 cm). English-made ell rules for buying continental cloth were therefore made about 27 inches (69 cm) long, while those for the home trade were 37 – 45 inches (94–114 cm) in length. Some ell rules have a variety of different ell lengths marked off.

Yardsticks and ell rules make attractive collectors' items, and can be found from the 17th century, although those not in museums are likely to be from the 18th and 19th centuries. Ell rules are usually made of boxwood, mahogany or pine, with brass ends, and sometimes brass inlay for the scale. Some may

be attractively inlaid or carved. Yardsticks, too, were generally made of wood, but examples exist of metal standard yardsticks, intended for use in manorial courts. These are made of brass, usually with a case, and will bear the name of the court, and probably the date and a royal cipher.

An early attempt to decimalise length measures produced the chain of 100 links. This was devised in 1620 by Edmund Gunter (1581–1626), the Gresham Professor of Astronomy at London. The length of a chain is 66 feet, equal to one-eightieth of a mile (1 mile = 5, 280 feet). This means that 10 square chains equal 1 acre. In the United States of America, the chain is the unit prescribed by law for surveying public lands.

Capacity Measures

The so-called 'Winchester Standard' (the association with Winchester dates from Saxon times, when it was the capital city of England) was established by Henry VII in 1497. The

Apothecary's reference standard measures, English, issued by the Board of Trade in 1878. The five glass goblets range from 1 fluid drachm to 3 fluid ounces, correct at 62° F. (SM)

measures were defined by the weight in troy ounces of their contents of wheat, by 'striked measure'; this means with the top levelled off by a spatula or levelling stick. The Winchester measures were declared standard for wheat, wine and ale: a pint = 12½ troy ounces of wheat; a quart (2 pints) = 25 troy ounces of wheat; a pottle (4 pints) = 50 troy ounces of wheat; a gallon (8 pints) = 100 troy ounces of wheat; and a bushel (64 pints) = 800 troy ounces of wheat.

The original Winchester standard measures for the gallon and the bushel, dating from the 15th century, and made of heavy bronze, can be seen in the Science Museum, London. Standard measures of this type were made until 1824, when the British Imperial measures were established. Standards, generally made of brass, may still be found dating from the 18th and early 19th centuries.

The other historic capacity measure was the wine gallon, which by custom was smaller than the Winchester gallon. Because this caused confusion, it was decided to define the wine gallon in 1707, under an Act of Queen Anne, as equivalent to 231 cubic inches. By comparison, the Winchester gallon was equivalent to 268 cubic inches, and the Imperial gallon of 1824 to 277 cubic inches. This is why different measures for wine gallons and ale gallons (ale and beer were measured by the Winchester standard) appear on antique measuring equipment. An interesting point is that the Winchester bushel and the Queen Anne wine gallon passed to the United States, and were legally adopted in 1836, where they remain the standards to this day, differing from the British Imperial measures that were brought in after 1824.

Antique capacity measures made for use in business and in the home, unlike the official standards, were generally made of ash, beech or oak. Measures for grain and other dry goods were made of thin, steamed bentwood, often banded with iron, and were the work of coopers. Measures for liquids, made by turners, were from hollowed-out solid wood. Alternatives were metal and horn. Almost all old measures are stamped with the crown and initials of the reigning monarch, and usually with a date. A 20th-century measure may also carry a number, which will be the reference number of the Weights and Measures office which verified it.

What we think of as liquid measures, the gill, pint, quart and gallon, were also, as mentioned, used for dry goods, such as grain, peas and beans. So the liquid and dry measures were more or less complementary.

Liquid Measure
4 gills/20 fl. oz = 1 pint
2 pints = 1 quart
4 quarts = 1 gallon

Beer Measure
4 ½ gallons = 1 pin
9 gallons = 1 firkin
18 gallons = 1 kilderkin
36 gallons = 1 barrel
54 gallons = 1 hogshead
72 gallons = 1 puncheon
108 gallons = 1 butt

Dry Measure
2 gallons = 1 peck
8 gallons = 1 bushel
64 gallons = 1 quarter

Wine Measure
42 gallons = 1 tierce
63 gallons = 1 hogshead
84 gallons = 1 puncheon
126 gallons = 1 pipe or butt
252 gallons = 1 tun

Makers

By the 18th century, scale and balance making was a specialist craft, as is indicated by the trade cards that may be found in some balance boxes. Among these are Henry Neale and William Brind of London, Thomas Beach of Birmingham, and Collet Frères of Paris. A skilled and important maker was R. B. Bate of London, who was a specialist metrologist, making hydrometers and saccharometers for the Customs and Excise, and the standard weights and measures after 1824 when Imperial standards were brought in. Another specialist firm of high reputation was Richard Vandome & Co., who made standards for the Bank of England and the East India Company. A name that appears on a number of standard measures of length and capacity, and also on many weights, is that of de Grave of London, who was working in the first part of the 19th century. After Charles de Grave died, his widow, Mary, continued to run the business, signing herself 'Mary de Grave, widow of Chas. de Grave'.

The leading British firm through the 19th century was W. & T. Avery, who took over Thomas Beach in 1817. On the death of William Avery in 1843, the firm was managed by his sons William and Thomas. In 1870, the West Bromwich Atlas Foundry in Birmingham was opened, and by 1885 it employed 770 people. The Soho foundry of James Watt & Co was acquired in 1895, to make Avery the largest scale-making company. In 1920, Avery took over De Grave, Short & Fanner, keeping the name until 1962.

MEDICAL INSTRUMENTS

T HE Greek physician Galen, in the second century AD, provided the foundations of the medical teaching and practice that prevailed virtually until the 18th century. He wrote over 500 treatises on anatomy and physiology, and set out the theory of the four humours that dominated all medical treatment throughout the Middle Ages. The humours were melancholy, phlegm, blood and choler, and imbalance between them in the human body was thought to be the fundamental cause of all disease. The physician's task was to restore the balance, and this was generally attempted by copious bleeding and changes of diet. Galen's anatomy, based on non-human material, was superseded, in the 16th century, by the work of Andreas Vesalius, whose superbly illustrated anatomical texts standardised the nomenclature of the human skeleton. The experimental studies of William Harvey founded the modern theory of blood circulation in 1628. Then, in the latter half of the 17th century, Marcello Malpighi used the newly invented microscope to observe the capillary circulation in the lung and bladder of a frog. The microscope was to become, in the 19th century, an important tool of medical diagnosis.

The interest in experimental philosophy that began in the late 1600s affected medicine, at first most noticeably through the teaching of Herman Boerhaave at Leiden University in the Netherlands, for his pupils went on to practise all over Europe. In the following centuries Britain became pre-eminent, with the work of James Simpson in pioneering the use of anaesthetics, and that of Joseph Lister in promoting antisepsis. Another important change during the 19th century was the reform of public health, particularly in the provision of sewage systems.

Surgery was practised from very early times, and is essentially practical. The name comes from the French *chirurgie* which is from the Greek, meaning 'hand-work'. Instruments were essential, and surgical instruments found in the ruins of the Italian city of Pompeii, which was wrecked by an eruption of Vesuvius in 79BC, are remarkably modern in appearance. But then the human body has not changed, and scalpels and saws are the constant stock-in-trade. Surgeons were needed on the battlefield, so it is hardly surprising that medieval makers of instruments were the armourers. New forms of wounds – gunshot rather than blade – required new methods.

The bow saw (like a modern hacksaw) and the tenon saw, which derive from Roman times, evolved according to the type of amputation.

The 17th century saw improvements to surgery along with new attitudes to science. The man known as the father of English surgery, Richard Wiseman, raised the status of his profession, as well as improving techniques. He was a royalist in the Civil War, and became immediate medical attendant to Prince Charles; he was appointed principal surgeon to the court of Charles II after the Restoration. He published works of surgery, and became Master of the Barber Surgeons' Company in the City of London.

Throughout Europe, surgery improved greatly during the 18th century, with qualified practitioners beginning to take over from barber-surgeons. Teaching colleges were set up in Paris, Vienna, Berlin and Edinburgh, while in London, there were private schools, notably that of William Hunter. Yet the Royal College of Surgeons of London was not founded until 1800. All this activity naturally generated a demand for instruments, and this was met by the cutlers. The trade in surgical instruments has always been independent of other forms of scientific equipment. Late Victorian and Edwardian trade catalogues, such as those produced by the firms Weiss and Down Bros., provide an interesting insight into the variations of surgical instruments.

Surgical Instruments

Surgical saws that have survived from the 17th century are often fine examples of craftsmanship, with handles of ivory, fishskin, or fine woods. Evidence of attention to technical requirements appears by the end of the 18th century, with lighter saws, the handles crosshatched in ebony for a firm grip.

The Napoleonic Wars created a demand for surgeons' kits; consequently, cases of instruments may be found from this time. Such kits are probably in red-lined mahogany boxes. By the early 20th century, amputation sets are in brassbound mahogany cases, and consist of a large saw (tenon saw type), a finger saw, Hey's skull saw, large and small amputation knives, scalpels, artery forceps, bullet forceps, catheters, gun-shot probe, ligature silk, needles and tourniquets. Sir William Hey designed his small-bladed skull saws in 1803 while working at the Leeds infirmary, and every kit has included one or more ever since. Large numbers of kits were sold to army and navy surgeons, and to those employed on passenger ships. Being the property of the surgeon himself, some will have survived to reach the collectors' market.

Trepanning is held to be even older than amputation. Neolithic skeletons have been found with holes bored in the skull, presumably done with flint knives. Traditional reasons for head-boring included the release of demons that caused headaches or epilepsy, and relief of a depression or fracture brought about by a blow. It is recorded that in the 16th century, William I, Prince of Orange, was trepanned 17 times for the relief of migraine.

There is often confusion in the name of an instrument. A trepan is a surgical instrument in the form of a crown saw, an iron cone with teeth filed into the lower rim. A trephine is also a crown saw, but represents a later form, after about 1630, which has a central pin to guide and steady the instrument. Trepanation is an operation with the trepan, and trephination with the trephine. Cased sets of trephines include a hand tool and four to six different-sized crown saws that are interchangeable. Tools either resemble a corkscrew, with a cross handle, or a carpenter's brace. Forceps, drills and other perforators will probably be included. Elevators are another necessary tool for lifting portions of fractured skull; they look rather like manicure implements. Most kits available today are 18th- and 19th-century, but 20th-century kits are also obtainable.

Blood-letting Instruments

Phlebotomy signifies blood-letting, literally from the Greek, vein-cutting. It is important in the four-humours theory of Galen, and it became a general all-embracing 'cure'. The lancet is the instrument used for blood-letting, and sets can often be very attractive, each blade protected in ivory or tortoiseshell, contained in silver or fishskin cases. The blade is double-sided, with a sharp point. A fleam is a lancet in which the pointed blade is at right-angles to the shaft, rather like a hatchet. The name can also refer to a lancet for bleeding horses. A scarificator is the name given to an instrument that makes several incisions at the same time. A small brass box holds four to twelve razor-sharp, curved blades that are retracted below the slotted surface of the instrument and then released by a trigger. The blades flash out to give a depth of cut governed by a turnscrew. (They must be handled with extreme caution.) Most scarificators found today are 19th-century, but they were still offered for sale in 1900. At that date a set of two, one with 12 blades and one with four

to be used on the temple, together with six cupping glasses, and a spirit lamp, were sold in a brass-bound mahogany case.

The cupping glass is a small glass bell-jar, which is placed over the wound caused by the lancet or scarificator to draw out and to collect the blood. Blood is drawn by heating the air in the glass before applying it to the skin. On cooling, a vacuum is produced, which readily sucks the blood from the wound. Cupping glasses may be found with a brass collar at the closed end and an attachment for an exhausting syringe. Here the air can be pumped out of the little bell-jar; this new technique of obtaining a vacuum is attributed to the experimental philosophers of about 1710.

Catheters

This is a thin tube (probably of silver) for passing into the bladder. It is used to remove urine from the bladder, or to apply solutions to alleviate strictures or ulcers. The Romans used such an instrument. Catheters for men normally come in cased sets of twelve of different gauge. They are about 12 inches (30 cm) long, with a curved end which is perforated. For women they are shorter, wider and straight.

Lithotomy Instruments

These are for removing stones which form in the bladder as a result of a poor and monotonous diet, and the operation to remove a stone is centuries-old. Many instruments were devised to grasp and extract the stone through an incision. In the first half of the 19th century tools were devised to crush the stone while still in the bladder.

Obstetric Instruments

Childbirth was largely left in the hands of female midwives until the mid-17th century. The earliest obstetric operation was craniotomy, the crushing of the skull of a dead foetus to make a still-birth possible and so save the mother's life. The first great pioneer of obstetrics was Ambroise Paré who, in the 16th century, founded a school for midwives in Paris. The earliest obstetrical forceps were used secretly in Britain by Peter Chamberlen at the beginning of the 17th century, and they were gradually improved in design and came into general use through the work of doctors such as William Smellie, in the 1740s and 1750s, who specialised in breach births, and lectured both the midwives and male colleagues, using living patients. It was, however, the 19th-century discovery of anaesthesia and the introduction of antiseptic practices that transformed the experience of childbirth, and greatly reduced infant mortality.

Post Mortem Instruments

These are similar to amputation sets but include a combined mallet and hatchet, plus a chisel.

Diagnostic Aids

Readily available diagnostic aids include the hammer, called a percussor (not always recognisable for what it is), and the ubiquitous stethoscope. In medical terms, percussion is the act of striking with one finger or with a small hammer against another finger placed on the surface of the body. The sound produced tells the doctor the physical state of that part of the body. The monaural stethoscope was invented by the French physician, René Laennec in 1816. His treatise of 1819 was not translated into English until 1825. About 12 inches (30 cm) long, with a wide rim at one end (for the ear), this instrument was made of fine woods, ivory, pewter or silver. The familiar binaural stethoscope was patented in 1855 by G. P. Caniman of New York, but did not immediately oust the single-tube type, which was sold in 1900 scarcely changed over the years, and is still used in maternity cases.

Medical Electricity

The interest aroused in the 18th century by frictional electricity has already been mentioned. It was generally believed that the administration of mild electric shocks would relieve illness and improve health. A medical electrical machine was patented in 1782 by Edward Nairne, and the popularity of the treatment continued into the present century, with the early type of machine superseded by the use of an induction coil, and later still by current electricity (**76**).

Medicine Chests

The apothecaries prepared and sold drugs and medicines, and the London Society of Apothecaries had the right to license medical practitioners. Physicians would need to have their own stock of preparations because of difficulties of travel in rural areas, and some larger households would also have their medicine chest (**74**). These can be attractive, with the finely made bottles in a red-lined mahogany chest, and may include a pestle and mortar, a simple balance with weights and measuring flasks. Some of the medicines for the household are likely to be emetic tartar, spirits of lavender, laudanum, milk of sulphur, bark, tincture of myrrh, basilicon, Rochelle salts, magnesia, spirits of hartshorn, pile ointment and liniment.

Savory & Moore is a Bond Street, London, firm that has provided medicines and drugs from the 1790s to the present day.

Dental Instruments

Dentistry was not a recognised branch of medicine until the mid-19th century. Although barber-surgeons would have been willing to attend to teeth, the minor operation of extracting teeth was an easy one, and fell into the province of fairground tooth-drawers and even blacksmiths. The medieval instrument was the 'pelican', named after its appearance. Its two parts could lock round a tooth and tighten as leverage was applied.

It was replaced by the tooth-key, invented in England in 1742, and called in Europe the *clef anglais*. Early ones had a handle and a shaft just like a key of the period; at the end was a claw projecting to one side, which could grip as the key was twisted. In the 19th century these were superseded by the dental forceps, which resembled nippers, with crooked and cleft blades. The forms varied for the left and right side of the mouth, for the different teeth, and for children and adults. Although superseded, the tooth-key was still offered for sale

in 1900, but with an ebony handle like a corkscrew. Treatment to the teeth, as opposed to removal, was rare in Britain until the Dental Society was founded in 1856. The bow-drill had been used in the 18th century, and George Washington's dentist, John Greenwood of Boston, used a spinning wheel to drive a drill. By the 1860s, the motorised dental drill had been introduced.

Sets of forceps and of drills exist. Included in them will be many other implements for special jobs, such as extracting roots, scaling, applying fillings, files and, of course, the dental mirror. The 1908 catalogue of the dental supplier, Claudius Ash, Sons & Co. Ltd, with branches all over Europe, extended to nearly 1,000 pages.

Ophthalmology

Care of the eyes did not become a specialised branch of medicine until after 1750. Spectacles, in the form of positive lenses to magnify, and so help older people to read, were in use from the 13th century. The lenses were made first by glass-workers, and then by optical instrument makers, but they were sold on a trial and error basis, and eye complaints were treated by any doctor. Cataract was the first condition to be relieved by an operation, followed by glaucoma in 1857. The ophthalmic surgeon's tools were more delicate versions of those used for general surgery, but eye-testing kits were made, and instruments such as the optometer and the ophthalmoscope were developed in the 1900s.

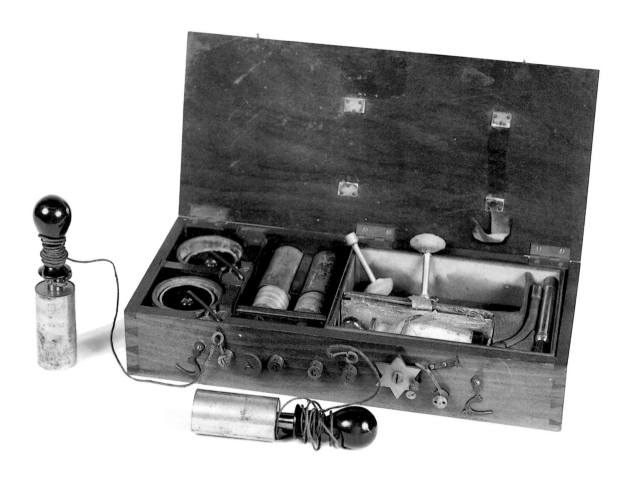

Electromedical induction apparatus, signed: RUHMKORFF *MÉCANICIEN*, 15 Rue des Maisons Sorbonne, Paris. By holding the brass parts on the handles, the patient receives a tingling shock. The batteries are mercury sulphate cells. Purchased in 1855. (TM)

PRACTICAL ADVICE

A GE AND RARITY are the most significant considerations in determining the value of an antique scientific instrument, but these are not the only factors; others to take into account are condition, completeness, historical significance and current tastes in collecting.

Dating

The simplest means of establishing the age of an instrument is to find a date inscribed on it. In a few cases, there is the risk of forgery, and, if a date only is found, it should be examined carefully for possible alteration, and for compatibility of style with the period. A date associated with the maker's signature is the most satisfactory form of dating.

When a signature and date appear on an instrument, it is interesting to consult a reliable reference work (see Bibliography) to discover any details that are known about the maker. In the Victorian period, and later, the name may well be that of a retailer rather than the maker, since quantity production began to get under way in the early 19th century, and even before. A maker's name, associated with a date in the 18th century or earlier, may endow the instrument with particular historical interest, and it would be worthwhile consulting a specialist museum (see Museums and Collections).

Many instruments bear the name of a maker, without any date. Here, a reference book can provide a date span. It is important to remember that a number of the leading firms of instrument makers were in business for long periods. The Adams family spanned nearly 100 years, and the firm of Dollond was operating from the 1750s to the early 20th century. So a microscope signed 'Dollond' should not immediately be thought to date from the 18th century; it may well be late 19th. The spelling 'Dolland' (with an a) should arouse immediate suspicion. The form of the signature, and the type of lettering used, can also give clues as to date to the expert. If the maker worked in London, the form of the address can help with dating. From 1767, numbers replaced signs which originally identified workshops and business premises. Benjamin Martin's address in 1767 was given for the first time as 171 Fleet Street. In 1857, 10 London postal districts were designated (SW, EC, etc). In 1912, numbers were added to the letters (SW10, EC4, etc).

Apart from an actual inscription on the instrument, there may be a trade label associated with it; for example, in the box of a microscope. These trade cards and labels are of considerable interest, often being embellished with a selection of the maker's stock-in-trade, and also with his shop sign. Edmund Culpeper, for instance, had his business at the sign of the Crossed Daggers, and this appears on his trade card.

Sometimes, other documents are associated with an instrument, such as instruction leaflets, or other advertising material; or perhaps some note or letter giving provenance. In certain cases, these can be of great interest, though there is scope for error and substitution. This again is a matter for a museum.

Because they provide a means of dating, signatures almost invariably add to the value of instruments. Certain signatures are particularly important, as belonging to makers of high repute; for example, James Short on telescopes, Edward Troughton on astronomical instruments, Ramsden on sextants, and Powell & Lealand, or Plössl of Vienna, on microscopes.

Dating an instrument without the aid of an inscribed date or maker's signature or label is a matter of studying style, materials and details of workmanship, and of knowing something of the history and development of the particular instrument in question. Careful examination of the photographs in this book, as well as visits to museum collections, will help to create the feel for different periods. The brief accounts of the history of a wide range of instruments given in the foregoing chapters will provide at least an introduction to the broad outlines of development.

Material

The expert's ability to ascribe an antique object to a particular period is built up from observation of all sorts of details: shape, decoration and such minutiae as screws, key escutcheons, handles, catches, hinges, box lining material, and so on. This kind of knowledge is largely the result of long experience, but some basic guidelines can be provided.

Many scientific instruments are made wholly or partly from wood, or else they have a container or carrying-case of wood. Broadly speaking, the use of mahogany indicates that the object was made later than the beginning of the Georgian period. Mahogany began to be imported into England from the West Indies at the beginning of the 18th century, and the tariff on it was greatly reduced in 1733, making its use common in fine cabinet-making. Prior to this, oak, walnut or fruit woods were used. But the arrival of mahogany, at first imported from Cuba and Jamaica, did not preclude the use of oak and other European woods. In the late 19th century, African mahogany came into extensive use; it is lighter in colour than the American wood.

It is almost impossible to date brass, even with modern laboratory techniques, because it is a variable alloy, mainly of copper and zinc, though other elements such as tin may be present in small quantities. With a brass instrument, methods of construction and design have to be relied upon to give clues as to date. Even with instruments made wholly of silver, it is rare to find any hallmark, however fine the quality. Silver is often tarnished black by sulphur in the atmosphere, and may not be readily recognised. But careful cleaning with non-abrasive material of a small portion will establish what the metal is. Sometimes a fine layer of gold has been applied over silver or copper. In old instruments, some of the gilding may have worn off, and any cleaning must be done with great care.

Pasteboard covered with various types of leather was used on telescope and microscope tubes until about 1750, and is also found on instrument boxes. The leather is frequently tooled or stamped in gold leaf. The tooling can provide clues as to country of origin and date, a technique resulting from careful study of the various designs. Shagreen is the name given to the skin of the sting-ray with scales ground flat, and the surface polished and dyed. This extremely hard material was used on microscope and telescope tubes until the late 18th century. It continued in use for boxes of various kinds, particularly for toilet sets, until the mid-20th century. In the 18th and early 19th centuries, sharkskin, invariably dyed black, and rough to the touch in one direction, was used to make cases and boxes for instruments.

It is possible to identify elephant ivory by the 'growth rings' observable in cross-section. Ivory is very hard and impervious, and is almost impossible to date. It was much used in the 18th century for microscope slides, and the screw barrels and handles of simple microscopes. In Victorian times, slides were often made of bone, which was cheap. Bone is not as white as ivory, and often has tiny dark flecks in the surface. Ivory is sometimes stained black in 18th-century instruments, and may be mistaken at first sight for ebony. Cheap wood, such as pine, was sometimes 'ebonised', or dyed black, again for cheapness, and used to manufacture the box-foot of an 18th-century microscope. In 1846, the process of 'vulcanising' was patented; this involved heat-treating rubber with sulphur to make it hard. This material was called either vulcanite or ebonite. Again, it may be mistaken for ebony, but close inspection will reveal it has no grain. With age, it goes matt in appearance.

The fabric with which boxes are lined can often provide a clue as to country of origin and date, although it is unwise to place too much reliance on this. Silk, especially watered silk, was often used in France; scarcely ever in England. A chamois-leather lining is often found in boxes of instruments made in Germany. Eighteenth-century English instruments most frequently had cases lined with green baize or green velvet. Blue and claret are often found in the mahogany boxes of Victorian and Edwardian instruments.

The style of lock and key on a case can often provide help in dating the instrument. Keys made in the 18th century may be distinguished by the thin, bow-shaped handle, developed from a crude twist of wire. Early keys usually have the leading edge of the bit cut away to pass the wards of the lock. After the lever lock was invented in 1778, the cuts are found on the bottom edge of the bit. If a box or case has a Bramah lock, this dates it to post-1784, when Joseph Bramah patented his design. Bramah locks are still made today; the key has a tiny, uncut bit, and is castellated at the end of the barrel.

Conservation

The basic principle in caring for antiques of all kinds is to interfere with them as little as possible until you have found out from an expert how to carry out repairs or renovations properly. Do not use abrasive cleaners; be particularly careful about cleaning brass and other metals too vigorously, as you will remove the original lacquer; re-lacquering will give too modern an appearance. It is wise to wear gloves when handling old brass instruments because of the risk of 'finger-etching'; this results from the salts in the skin reacting with the metal, and can cause lasting damage. Wooden instruments, or their boxes, that have been kept in a dry, heated atmosphere often come unglued. The old glue should then be carefully removed, and fresh animal or fish glue used in its place. The common sticky tapes should never be used for temporary repairs, as they can remove surfaces, and will eventually degrade.

BIBLIOGRAPHY

This bibliography is intended to provide an initial guide to books on the history of scientific instruments that place special emphasis on the instruments themselves. Many of these will have a bibliography relating to the subject of the particular book. It is not appropriate here to cite papers published in learned journals. The libraries in the museums listed in the following chapter will be able to provide information on makers and on instruments. For anyone interested in old scientific instruments, it is worth joining the Scientific Instrument Society, which publishes a substantial quarterly *Bulletin*, and occasional Monographs: SIS Executive Officer, 31 High Street, Stanford in the Vale, Oxon SN7 8LH, UK.

Andersen, Hemming, *Historic Scientific Instruments in Denmark*, Royal Danish Academy of Sciences and Letters, Copenhagen, 1995.

Anderson, R. G. W., Bennett, J. A., and Ryan, W. F., eds, *Making Instruments Count: Essays on Historical Scientific Instruments presented to Gerard L'Estrange Turner*, Variorum, Aldershot, 1993.

Anderson, R. G. W., Burnett, J., and Gee, B., *Handlist of Scientific Instrument-Makers' Trade Catalogues 1600-1914*, National Museums of Scotland Information Series No. 8, Edinburgh, 1990.

Banfield, Edwin, *Antique Barometers: An Illustrated Survey*, Wayland Publications, Hereford, 1976.

—, *Barometer Makers and Retailers 1660–1900*, Baros Books, Trowbridge, 1991.

Bedini, Silvio A., *Early American Scientific Instruments and Their Makers*, Smithsonian Institution, Washington, DC, 1964; reprinted by Landmark Enterprises, Rancho Cordova, 1986.

—, *Thinkers and Tinkers: Early American Men of Science*, Charles Scribner's Sons, New York, 1975.

—, *Science and Instruments in Seventeenth-Century Italy*, Variorum, Aldershot, 1994.

Bennett, J.A., *The Divided Circle: A History of Instruments for Astronomy Navigation and Surveying*, Oxford, 1987.

Bennion, Elisabeth, *Antique Medical Instruments*, Sotheby Parke Bernet, London, 1980.

—, *Antique Dental Instruments*, Philip Wilson Publishers, London, 1986.

—, *Antique Hearing Devices*, Vernier Press, London and Brighton, 1994.

Biggs, Norman, *English Weights: An Illustrated Survey*, White House Publications, Llanfyllin, Powys, 1992.

Bion, Nicolas, *The Construction and Principal Uses of Mathematical Instruments* (translated by Edmund Stone, London, 1758), reprinted Holland Press, London, 1972.

Bolle, Bert, *Barometers*, Argus Books Ltd, Watford, 1982.

Brachner, Alto, ed., *G.F. Brander 1713–1783: Wissenschaftliche Instrumente aus seiner Werkstatt*, Deutsches Museum, Munich, 1983.

Brachner, Alto, *Mit den Wellen des Lichts: Ursprünge und Entwicklung der Optik in süddeutschen Raum*, Olzog Verlag, Munich, 1987.

Bradbury, S., *The Evolution of the Microscope*, Pergamon Press, Oxford, 1967.

Brenni, Paolo, *Museo di Storia della Scienza: Catalogue of Mechanical Instruments*, Museo di Storia della Scienza, Florence, 1993.

Brieux, Alain, ed., *Catalogue du Maison Nachet, 1856–1910*, Editions Brieux, Paris, 1979.

Brown, Joyce, *Mathematical Instrument-Makers in the Grocers' Company 1688–1800*, The Science Museum, London, 1979.

Bruyns, W. F. J. Mörzer, *The Cross-Staff: History and Development of a Navigational Instrument*, Scheepvaart Museum, Amsterdam, 1994.

Burnett, J. E., and Morrison-Low, A. D., '*Vulgar & Mechanick': The Scientific Instrument Trade in Ireland 1650–1921*, National Museums of Scotland and Royal Dublin Society, Edinburgh and Dublin, 1989.

Calvert, H. R., *Scientific Trade Cards in the Science Museum Collection*, HMSO, London, 1971.

Chapman, Allan, *Dividing the Circle: The Development of Critical Angular Measurement in Astronomy 1500–1850*, second edition, John Wiley & Sons, Chichester, 1995.

—, *Astronomical Instruments and Their Users*, Variorum, Aldershot, 1996.

Chew, V. K., *Talking Machines*, The Science Museum, London, 1981.

Clarke, T. N., Morrison-Low, A. D., and Simpson, A. D. C., *Brass and Glass: Scientific Instrument Making Workshops in Scotland*, National Museums of Scotland, Edinburgh, 1989.

Clifton, Gloria, *Directory of British Scientific Instrument Makers 1550–1851*, Zwemmer, London, 1995.

Conner, R. D., *The Weights and Measures of England*, HMSO, London, 1987.

Crawforth, M. A., *Weighing Coins: English Folding Gold Balances of the 18th and 19th Centuries*, Cape Horn Trading Co., London, 1979.

Crompton, Dennis, Henry, D., and Herbert, S., eds, *Magic Images: The Art of Hand-Painted and Photographic Lantern Slides*, The Magic Lantern Society of Great Britain, London, 1990.

Curtis, Tony, ed., *The Lyle Official Antiques Review*, annually, Lyle Publications, Galashiels.

De Clercq, P. R., *At the Sign of the Oriental Lamp: The Musschenbroek Workshop in Leiden, 1660–1750*, Erasmus Publishing, Rotterdam, 1997.

Daumas, Maurice, *Scientific Instruments of the 17th and 18th Centuries and their Makers*, Batsford, London, 1972.

Davis, Audrey B., and Dreyfuss, Mark S., *The Finest Instruments Ever Made: A Bibliography of Medical, Dental, Optical, and Pharmaceutical Company Trade Literature; 1700–1939*, Medical History Publishing Associates, Arlington, Mass., 1986.

Edlin, Herbert L., *What wood is that?: A Manual of Wood Identification* (with 40 actual wood specimens), Stobart & Son Ltd, London, 1977.

Gillispie, C. C., ed., *Dictionary of Scientific Biography*, 15 vols, Charles Scribner's Sons, New York, 1970–80.

Goodison, Nicholas, *English Barometers 1680–1860: A History of Domestic Barometers and their Makers and Retailers*, second edn, Antique Collectors' Club, Woodbridge, 1977.

Gouk, Penelope, *The Ivory Sundials of Nuremberg 1500–1700*, Whipple Museum of the History of Science, Cambridge, 1988.

Hackmann, W. D., *Electricity from Glass: The History of the Frictional Electrical Machine 1600–1850*, Sijthoff & Noordhoff, Alphen aan den Rijn, 1978.

— *Museo di Storia della Scienza: Catalogue of Pneumatical, Magnetical and Electrical Instruments*, Museo di Storia della Scienza, Florence, 1995.

Hambly, Maya, *Drawing Instruments 1580–1980*, Sotheby's Publications, London, 1988.

Hammond, J. H., and Austin, J., *The Camera Lucida in Art and Science*, Adam Hilger, Bristol, 1987.

Holbrook, Mary, *Science Preserved: A Directory of Scientific Instruments in Collections in the United Kingdom and Eire*, HMSO, London, 1992.

Horský, Zdeněk, and Škopová, Otilie, *Astronomy Gnomonics: A Catalogue of Instruments of the 15th to* the 19th Centuries in the Collections of the National Technical Museum, Prague, National Museum, Prague, 1968.

King, Henry C., and Millburn, J. R., *Geared to the Stars: The Evolution of Planetariums, Orreries and Astronomical Clocks*, Toronto University Press, Toronto; Adam Hilger, Bristol, 1979.

Kisch, Bruno, *Scales & Weights: A Historical Outline*, Yale University Press, Newhaven, Conn., 1965.

Lloyd, Steven A., *Ivory Diptych Sundials 1570–1750, The Collection of Historical Scientific Instruments, Harvard University*, Harvard University Press, Cambridge, Mass., 1992.

Maistrov, L. E., *Pribory i instrumenty istoricheskogo znacheniya: Nauchnyi probory* [Apparatus and instruments of historic importance: Scientific apparatus], Nauka, Moscow, 1968.

McConnell, Anita, *Geophysics & Geomagnetism: Catalogue of the Science Museum Collection*, HMSO, London, 1986.

—, *Instrument Makers to the World: A History of Cooke, Troughton & Simms*, William Sessions, York, 1992.

—, *R. B. Bate, of the Poultry 1782–1847: The Life and Times of a Scientific Instrument-Maker*, Scientific Instrument Society, London, 1993.

Michel, Henri, *Scientific Instruments in Art and History*, Barrie & Rockliff, London, 1967. [Also published in French and German.]

Middleton, W. E. Knowles, *Catalog of Meteorological Instruments in the Museum of History and Technology*, Smithsonian Institution Press, Washington, DC, 1969.

Miniati, Mara, ed., *Museo di Storia della Scienza: Catalogo*, Museo di Storia della Scienza, Florence, 1991.

Minow, Helmut, *Historical Surveying Instruments: List of Collections in Europe*, second edn, Verlag Chmielorz GmbH, Wiesbaden, 1990. [Published in three languages.]

Mollan, Charles, *Irish National Inventory of Historic Scientific Instruments*, Samton Ltd, Dublin, 1995.

Mollan, Charles, and Upton, John, *St Patrick's College, Maynooth: The Scientific Apparatus of Nicholas Callan and other Historic Instruments*, St Patrick's College, Maynooth, 1994.

Mooij, J., *Instrumenten, wetenschap en samenleving: Geschiedenis van de instrumentenfabricage en -handel in Nederland 1840–1940*, Het Instrument, Soest, 1988.

Morton, A. Q., and Wess, J. A., *Public & Private Science: The King George III Collection*, Science Museum and Oxford University Press, London and Oxford, 1993.

Multhauf, R. P. and Good, G., *A Brief History of Geomagnetism and A Catalog of the Collections of the National Museum of American History*, Smithsonian Institution Press, Washington, DC, 1987.

National Maritime Museum, Greenwich, *The Planispheric Astrolabe*, HMSO, London, 1976; revised 1979.

Pullan, J. M., *The History of the Abacus*, Hutchinson, London, 1969.

Randier, Jean, *Nautical Antiques for the Collector*, English translation, Barrie & Jenkins, London, 1976. [Also in French and German.]

Richeson, A. W., *English Land Measuring to 1800: Instruments and Practice*, MIT Press, Cambridge, Mass., 1966.

Rohr, René R. J., *Sundials: History, Theory, and Practice*, University of Toronto Press, Toronto, 1970. [Also in French.]

Rooseboom, Maria, *Bijdrage tot de Geschiedenis der Instrumentmakerskunst in de noordelijke Nederlanden tot omstreeks 1840*, Museum voor de Geschiedenis der Natuurwetenschappen, Leiden, 1950.

Saunders, Harold N., *All the Astrolabes*, Senecio Publishing Co., Oxford, 1984.

Savage, George, *The Art and Antique Restorers' Handbook: A Dictionary of Materials and Processes used in the Restoration and Preservation of all kinds of Works of Art*, revised edn, Barrie & Jenkins, London, 1976.

Stimson, A. N., *The Mariner's Astrolabe: A Survey of Known Surviving Sea Astrolabes*, HES Publishers, Utrecht, 1988.

Stock, J. T., *Development of the Chemical Balance: A Science Museum Survey*, HMSO, London, 1969.

Stock, J. T., and Vaughan, Denys, *The Development of Instruments to Measure Electric Current*, HMSO, London, 1983.

Taylor, E. G. R., *The Mathematical Practitioners of Tudor and Stuart England*, Cambridge University Press, Cambridge, 1954.

—, *The Mathematical Practitioners of Hanoverian England*, 1714–1840, Cambridge University Press, Cambridge, 1966.

Thomas, D. B., *The Science Museum Photography Collection*, HMSO, London, 1969.

Turner, Anthony, *Early Scientific Instruments: Europe 1400–1800*, Sotheby's Publications, London, 1987.

—, *Of Time and Measurement: Studies in the History of Horology and Fine Technology*, Variorum, Aldershot, 1993.

Turner, G. L'E., *Descriptive Catalogue of Van Marum's Instruments in Teyler's Museum Haarlem* (Martinus van Marum: Life and Work, Volume iv, Part ii), Noordhoff for Hollandsche Maatschappij der Wetenschappen, Leiden, 1973.

—, *Collecting Microscopes*, Studio Vista, London, 1980. [Also in German, Dutch and Italian.]

—, *Nineteenth-Century Scientific Instruments*, Philip Wilson Publishers, London, and University of California Press, Berkeley, 1983).

—, *The Great Age of the Microscope: The Collection of the Royal Microscopical Society through 150 Years*, Adam Hilger, Bristol, 1989.

—, *Scientific Instruments and Experimental Philosophy 1550-1850*, Variorum, Aldershot, 1990.

—, *Museo di Storia della Scienza: Catalogue of Microscopes*, Istituto e Museo di Storia della Scienza, Florence, 1991.

—, *Gli Strumenti: Storia della Scienza*, vol. 1, General Editor Paolo Galluzzi, Giulio Einaudi Editore, Turin, 1991.

—, *The Practice of Science in the Nineteenth Century: Teaching and Research Apparatus in the Teyler Museum*, The Teyler Museum, Haarlem, 1996.

Ward, F. A. B., *A Catalogue of European Scientific Instruments in the Department of Medieval and Later Antiquities of the British Museum*, British Museum Publications, London, 1981.

Waters, David W., *The Art of Navigation in England in Elizabethan and Early Stuart Times*, Hollis and Carter, London, 1958; reprinted 1979.

Waugh, Albert E., *Sundials: Their Theory and Construction*, Dover Publications, New York, 1973.

Wheatland, David P., *The Apparatus of Science at Harvard, 1765–1800*, Harvard University Press, Cambridge, Mass., 1968.

Zinner, Ernst, *Deutsche und niederländische astronomische Instrumente des 11.–18. Jahrhunderts*, revised second edn, Beck'che Verlag, Munich, 1979.

MUSEUMS AND COLLECTIONS

A comprehensive guide to world museums is *The Directory of Museums* by Kenneth Hudson and Ann Nicholls, Macmillan, London, third edition 1985. A more recent world listing is *Museums of the World*, edited by Elisabeth Richter, Saur Verlag, Munich, fifth edition 1995. An essential guide to Britain is *Museums and Galleries in Great Britain and Ireland*, issued annually by ABC Historic Publications, Dunstable, Bedfordshire. For West Germany, there is the well-illustrated, 800-page book edited by Klemens Mörmann, *Der deutsche Museums Führer in Farbe*, Wolfgang Kruger Verlag, Frankfurt-am-Main, second edition 1983.

Austria
Museum of Industry and Technology, Mariahilfferstrasse 212, A-1140 Vienna 14

Belgium
Museum voor de Geschiedenis van de Wetenschappen, Gebouw S 30, Krijkslaan 281, B-9000 Gent

Canada
National Museum of Science and Technology, P.O. Box 9724, Ottawa, Ontario K1G 5A3

Czech Republic
National Technical Museum, Kostelni 42, 170 78 Prague 7

Denmark
The Steno Museum (The Danish Museum for the History of Science), C.F. Møllers Allé, University Park, DK-8000 Aarhus C

The Danish Technical Museum, Nordre Strandvej 23, DK-3000 Helsingør

The A.W. Hauch Collection of Physical Instruments, Sorø Akademi, 'Vænget', Søgade 17, DK-4180 Sorø

France
Conservatoire National des Arts et Métiers, Musée National des Techniques, 292, rue Saint-Martin, 75003 Paris

Musée de l'Observatoire, 61 Avenue de l'Observatoire, 75014 Paris

Germany
Deutsches Museum, Museuminsel 1, 8036 Munich

Germanisches Nationalmuseum, Kornmarkt 1, 8500 Nuremberg

Hessisches Landesmuseum, Bruder Grimm Platz 5, 3500 Kassel 1

Staatlischer Mathematisch-Physikalischer Salon in Zwinger, 01067 Dresden

Great Britain
The British Museum, Department of Medieval and later Antiquities, London WC1B 3DG

The Science Museum, South Kensington, London SW7 2DD

The Old Royal Observatory, National Maritime Museum, Greenwich, London SE10 9NF

Museum of the History of Science, Broad Street,
    Oxford OX1 3AZ
Whipple Museum of the History of Science, Free
    School Lane, Cambridge CB2 3RH
North Western Museum of Science and Industry,
    97 Grosvenor Street, Manchester M1 7HF
The Royal Scottish Museum, Chambers Street,
    Edinburgh EH1 1JF

India

Birla Industrial and Technological Museum, 19A
    Gurusaday Road, Calcutta 700019

Ireland

National Museum, Kildare Street, Dublin 2
St Patrick's College, Maynooth, Co. Kildare

Italy

Museo di Storia della Scienza, Piazza dei Giudici 1,
    I-50122 Florence

Netherlands

Museum Boerhaave, Lange St Agnietenstraat 10,
    2312 WC Leiden
The Teyler Museum, Spaarne 16, 2011 CH
    Haarlem
Universiteits Museum, Lange Nieuwstraat 106,
    3501 AB Utrecht

Russia

M. V. Lomonosov Museum, Universitetskaya Nab.,
    St Petersburg
Polytechnic Museum, 3 Novaya Ploshad, Moscow

Spain

Museo nacional de Ciencia y Tecnologia, Paseo de
    las Delicias 61, 28045 Madrid
Museo Naval, Paseo del Prado 5, 28071 Madrid

Sweden

Tekniska Museet, Museivagen 7, 115 27
Stockholm

Switzerland

Musée de l'Histoire des Sciences, 128 rue de
    Lausanne, 1202 Geneva

United States of America

National Museum of American History,
    Smithsonian Institution, Washington, D.C.
    20560
Historic Collections, The Adler Planetarium, 1300
    South Lake Shore Drive, Chicago, Illinois
    60605
Collection of Historical Scientific Instruments,
    Science Center, Harvard University,
    Cambridge, Massachusetts 02138

INDEX